Recent Trends of Mobile Collaborative
Augmented Reality Systems

Leila Alem · Weidong Huang
Editors

Recent Trends of Mobile Collaborative Augmented Reality Systems

 Springer

Editors
Leila Alem
CSIRO ICT Centre
Cnr Vimiera and Pembroke Roads
Marsfield, NSW 2122
Australia
Leila.Alem@csiro.au

Weidong Huang
CSIRO ICT Centre
Cnr Vimiera and Pembroke Roads
Marsfield, NSW 2122
Australia
Tony.Huang@csiro.au

ISBN 978-1-4899-9986-3   ISBN 978-1-4419-9845-3 (eBook)
DOI 10.1007/978-1-4419-9845-3
Springer New York Dordrecht Heidelberg London

Printed on acid-free paper

Springer is part of Springer Science+Business Media (www.springer.com)

Preface

Augmented reality (AR) is a direct or indirect view of real world scenes in which physical objects are annotated with, or overlaid by computer generated digital information. The past two decades have seen a fast growing body of research and development dedicated to techniques and technologies for AR. In particular, due to the recent advances in mobile devices and networking technologies, the use of mobile collaborative augmented reality (MCAR) has expanded rapidly. Although there is still a long way for MCAR systems to become commonplace, successful applications have been developed in a range of fields, for example computer-supported collaborative learning, entertainment, tourism and collaborative architectural design. An overview of recent trends and developments in this rapidly advancing technology is needed. This book is set out to:

- Provide a historical overview of previous MCAR systems
- Present case studies of latest developments in current MCAR systems
- Cover latest technologies and system architectures used in MCAR systems

This book includes 13 chapters. The first two chapters of this book are invited contributions from established researchers in the field. The remaining chapters are extended versions of papers presented in the 2010 international workshop on mobile collaborative augmented reality (MCAR 2010). We briefly introduce these chapters as follows.

In chapter 1, Billinghurst and Thomas provide an overview of current state-of-art in MCAR. The authors first introduce a set of technologies that are required for MCAR. Then examples of recent MCAR systems are presented. The chapter finishes with an insightful look into requirements and directions of future MCAR.

In chapter 2, Perey et al. discuss the needs, approaches, issues and directions of standardization of AR related applications and services. This discussion includes guiding principles of an open AR industry, AR requirements and use cases, approaches to the AR standards challenge and content-related standards. The current state of mobile AR standards is also introduced and analysed.

In chapter 3, Yew et al. propose a system framework called SmARt World which is to support various mobile collaborative applications in indoor environments.

This system has a three-layered architecture – physical layer, middle layer and AR layer. The initial prototype has been implemented and the tests of it show that it is low-cost, user friendly and suitable for many applications.

In chapter 4, Hoang and Thomas present research directions motived by the problem of action at a distance in mobile augmented reality. The discussion is based on the authors' augmented viewport technique. Current research challenges are identified, which include the utilization of various types of remote cameras, collaboration features, better visualization of the cameras' views, precision by snapping and improved input devices.

In chapter 5, Webel et al. present a series of analytic results of interdisciplinary research. Based on previous research and experiments performed in cooperation with human factors scientists, improvement of Augmented Reality based training of skills is analysed and recommendations for the design of Augmented Reality based training systems are proposed. These recommendations include visual aids, elaborated knowledge, passive learning parts and haptic hints.

In chapter 6, Vico et al. describe a taxonomy for classifying types of applications involving mobile AR-based collaboration. The authors propose that experiences can be classified according to the type of content generated and then give examples of how current mobile AR applications would be classified. Some possible use cases of the taxonomy and future research are provided.

In chapter 7, Gu et al. describe the development of a mobile AR collaborative game called AR Fighter. The structure and features of this game's prototype are introduced. In this prototype, the authors present a concept of game playing: 2 players can play an AR game without any onlookers interfering.

In chapter 8, Alem et al. present a user study of an augmented reality mobile game called Greenet. This game allows players to learn about recycling by practicing the act of recycling using a mobile phone. The study compares three different ways of playing the game and the results suggest that competitive/collaborative mobile phone based games provide a promising platform for persuasion.

In chapter 9, Gu et al. present a game called AR-Sumo. This game is a mobile collaborative augmented reality network service for educational and entertainment purposes. AR-Sumo provides a shared virtual space for multiple users to interact at the same time. It involves visualization of augmented physical phenomena on a fiducial marker and enables learners to view the physical effects of varying gravities and frictions in a 3D virtual space.

In chapter 10, Wang et al. propose a multi-user guide system for Yuanmingyuan Garden. The system integrates real environments and virtual scenes through entertainment and gaming in mobile phones. Using this system, visitors are able to tour the garden's historical sites and experience the excitements of an AR based game through various novel ways of interaction provided.

In chapter 11, Alem et al. present a gesture based mobile AR system for supporting remote collaboration called HandsOnVideo. The system is developed following a participatory design approach. It can be used for scenarios in which a remote helper guides a mobile worker in performing tasks that require the manipulation of physical objects, such as maintaining a piece of equipment and performing an assembly task.

In chapter 12, White and Feiner present a system for dynamic, abstract audio representations in mobile augmented reality called SoundSight. This system uses the Skype Internet telephony API to support wireless conferencing and provides visual representations of audio, allowing users to "see" the sounds. Initial user experience of the system indicates that visual representations of audio can help to promote presence and identify audio sources.

In chapter 13, Zhou et al. review Spatial Augmented Reality (SAR) techniques. Advantages and problems of SAR in presenting digital information to users are summarised. The authors also present a concept of portable collaborative SAR. This concept is then applied in a case study of an industrial quality assurance scenario to show its effectiveness.

In summary, the research topics presented in this book are diverse and multidisciplinary. These topics highlight recent trends and developments in MCAR. We hope that this book is useful for a professional audience composed of practitioners and researchers working in the field of augmented reality and human-computer interaction. Advanced-level students in computer science and electrical engineering focused on these topics should also find this book useful as a secondary text or reference.

We wish to express our gratitude to Professor Mark Billinghurst and Professor Bruce H. Thomas for their help and support throughout this project. We also would like to thank the members of the international editorial board for their reviews and all authors for their contributions to the book. Last but not least, we would like to thank Susan Lagerstrom-Fife and Jennifer Maurer at Springer USA for their assistance in editing this book.

Leila Alem
Weidong Huang

Contents

Contributors

Leila Alem CSIRO ICT Centre, PO Box 76, Epping NSW 1710, Australia

Peta Ashworth CESRE Division, PO Box 883, Kenmore QLD 4069, Australia

Mark Billinghurst HIT Lab NZ, University of Canterbury, Private Bag 4800 Christchurch 8140, New Zealand

Ulrich Bockholt Fraunhofer Institute for Computer Graphics Research IGD, Fraunhoferstr. 5, 64283 Darmstadt, Germany

Leanne Chang Communications and New Media Programme, Faculty of Arts & Social Sciences, Blk AS6, #03-41, 11 Computing Drive, National University of Singapore

Henry Been-Lirn Duh Department of Electrical and Computer Engineering, 21 Heng Mui Keng Terrace, #02-02-09, National University of Singapore

Steven Feiner Columbia University, 500 W. 120th St., 450 CS Building, New York, NY 10027

Timo Engelke Fraunhofer Institute for Computer Graphics Research IGD, Fraunhoferstr. 5, 64283 Darmstadt, Germany

David Furio Instituto Universitario de Automática e Informática Industrial, Universidad Politécnica de Valencia, Camino de Vera, s/n. 46022, Valencia, Spain

Nirit Gavish Technion Israel Institute of Technology, Technion City, Haifa 32000, Israel

Jian Gu KEIO-NUS CUTE center, Interactive & Digital Media Institute, 21 Heng Mui Keng Terrace, #02-02-09, National University of Singapore

Yuan Xun Gu Department of Electrical and Computer Engineering, 21 Heng Mui Keng Terrace, #02-02-09, National University of Singapore

Thuong N. Hoang Wearable Computer Lab, School of Computer and
Information Science, University of South Australia, Mawson Lakes Campus,
1 Mawson Lakes Boulevard, Mawson Lakes, SA 5010, Australia

Weidong Huang CSIRO ICT Centre, PO Box 76, Epping NSW 1710, Australia

Xia Jia ZTE Corporation, Nanjing, 518057, China

Carmen Juan Instituto Universitario de Automática e Informática Industrial,
Universidad Politécnica de Valencia, Camino de Vera, s/n. 46022, Valencia, Spain

Shintaro Kitazawa KEIO-NUS CUTE center, Interactive & Digital Media
Institute, 21 Heng Mui Keng Terrace, #02-02-09, National University
of Singapore

Ivan Lee School of Computer and Information Science, University of South
Australia, Australia

Nai Li Communications and New Media Programme, Faculty of Arts & Social
Sciences, Blk AS6, #03-41, 11 Computing Drive, National University
of Singapore

Yue Liu School of Optics and Electronics, Beijing Institute of Technology,
Beijing, 100081, China

Roland Menassa GM Michigan, North America

A.Y.C. Nee Department of Mechanical Engineering, Augmented Reality Lab,
National University of Singapore, EA 02-08, National University of Singapore,
117576, Singapore

S.K. Ong Department of Mechanical Engineering, Augmented Reality Lab,
National University of Singapore, EA 02-08, National University of Singapore,
117576, Singapore

Christine Perey Spime Wrangler, PEREY Research & Consulting

Carl Reed Open Geospatial Consortium

Joaquín Salvachúa Rodríguez Departamento de Ingeniería de Sistemas
Telemáticos, Escuela Técnica Superior de Ingenieros de Telecomunicación,
Universidad Politécnica de Madrid, Avenida Complutense 30, "Ciudad
Universitaria", 28040, Madrid, Spain

Andrew Sansome GM Holden Ltd, Victoria, Australia

Franco Tecchia PERCRO Laboratory - Scuola Superiore Sant'Anna,
Viale R. Piaggio 34, 56025 Pontedera (Pisa), Italy

Bruce H. Thomas Wearable Computer Lab, School of Computer and
Information Science, University of South Australia, Mawson Lakes Campus,
1 Mawson Lakes Boulevard, Mawson Lakes, SA 5010, Australia

Iván Martínez Toro Departamento de Ingeniería de Sistemas Telemáticos, Escuela Técnica Superior de Ingenieros de Telecomunicación, Universidad Politécnica de Madrid, Avenida Complutense 30, "Ciudad Universitaria", 28040, Madrid, Spain

Daniel Gallego Vico Departamento de Ingeniería de Sistemas Telemáticos, Escuela Técnica Superior de Ingenieros de Telecomunicación, Universidad Politécnica de Madrid, Avenida Complutense 30, "Ciudad Universitaria", 28040, Madrid, Spain

Yongtian Wang School of Optics and Electronics, Beijing Institute of Technology, Beijing 100081, China

Sabine Webel Fraunhofer Institute for Computer Graphics Research IGD, Fraunhoferstr. 5, 64283 Darmstadt, Germany

Sean White Columbia University & Nokia Research Center, 2400 Broadway, Suite D500, Santa Monica, CA 90404

Jian Yang School of Optics and Electronics, Beijing Institute of Technology, Beijing 100081, China

A.W.W. Yew Department of Mechanical Engineering, Augmented Reality Lab, National University of Singapore, EA 02-08, National University of Singapore, 117576, Singapore

Liangliang Zhai School of Optics and Electronics, Beijing Institute of Technology, Beijing 100081, China

Zhipeng Zhong School of Optics and Electronics, Beijing Institute of Technology, Beijing 100081, China

Jianlong Zhou School of Computer and Information Science, University of South Australia, Australia

Mobile Collaborative Augmented Reality

Mark Billinghurst and Bruce H. Thomas

Abstract This chapter provides an overview of the concept of Mobile Collaborative Augmented Reality (MCAR). An introduction to augmented reality is firstly provided which gives an insight into the requirements of mobile augmented reality (some-times referred to as handheld augmented reality). A set of current MCAR systems are examined to provide context of the current state of the research. The chapter finishes with a look into the requirements and future of MCAR.

1 Introduction

Augmented Reality (AR) is a technology that allows interactive three-dimensional virtual imagery to be overlaid on the real world. First developed over forty years ago [1], applications of Augmented Reality have been employed in many domains such as education [2], engineering [3] and entertainment [4]. For example, mechanics are able to see virtual instructions appearing over real engines giving step by step maintenance instructions [5], and gamers can see virtual monsters appearing over real playing cards and fighting with each other when they are placed side by side [6]. Azuma provides a detailed review of current and past AR technology [7, 8].

Figure 1 shows a typical AR interface, in this case the user's view through a head mounted display (HMD) while looking down a street. The physical world is the building in the background, and a virtual road, street lamps, and houses appear overlaid on the real world in front of it. This particular AR application lets the user enhance the landscape outside their building with the addition of a virtual road, houses and a set of street lamps that can be walked around. The virtual objects are

M. Billinghurst (✉)
HIT Lab NZ, University of Canterbury, Private Bag 4800 Christchurch, 8140, New Zealand
e-mail: mark.billinghurst@canterbury.ac.nz

L. Alem and W. Huang (eds.), *Recent Trends of Mobile Collaborative Augmented Reality Systems*, DOI 10.1007/978-1-4419-9845-3_1,
© Springer Science+Business Media, LLC 2011

Fig. 1 Users view of the Real World

Fig. 2 Tinmith Hardware

registered to the physical world and appear at – or standards based - content, platforms or viewing applications. It is a field of technology silos and, consequently fragmented markets.

For mobile collaborative AR, the needs for standards are compounded by the fact that the content of shared interest must travel over a communications "bridge" which is, itself, established between end points between and through servers, client devices and across networks. The more interoperable the components of the end-to-end system are, the less the fixed locations in it. The Tinmith [9] AR wearable computing hardware is an example system that supports this form of AR, see Figure 2.

Figure 1 shows that a major benefit of AR is the viewing of information that is location based and registered to physical objects. Basic AR systems provide information about the physical world, and let users view that information in any setting. For example, a classic use is to visualize a proposed architectural structure in the context of existing buildings or at a particular physical location. The ability to walk in and around the virtual structure lets users experience its size, shape, and feel in a first-person perspective and fosters a more emotional engagement.

Recently, mobile phones have become as powerful as the desktop computers from a decade earlier, and so mobile augmented reality has become possible. Modern smart phones combine fast CPUs with graphics hardware, large screens, high resolution cameras and sensors such as GPS, compass and gyroscopes. This makes them an ideal platform for Augmented Reality. Henrysson [10], Wagner [11] and others have shown how computer vision based AR applications can be delivered on mobile phones, while commercial systems such as Layar[1], Wikitude[2], and Junaio[3] use GPS and compass sensor data to support outdoor AR experiences.

Phones also have powerful communication hardware, both cellular and wireless networking, and can be used for collaboration. So, for the first time, consumers have in their hands hardware that can provide a collaborative AR experience [12]. A Mobile Collaborative Augmented Reality (MCAR) application is one that allows several people to share an AR experience using their mobile devices [13]. The AR content could be shared among face to face or remote users, and at the same time (synchronous collaboration) or at different times (asynchronous collaboration).

1.1   Core Mobile AR Technology

In order to deliver a MCAR experience there are several core pieces of technology that must be used, including:

- *Mobile Processor*: Central Processing Unit (CPU) for processing user input, video images and running any application simulations.
- *Graphics Hardware*: Graphical Processing Unit (GPU) system for generating virtual images.
- *Camera*: Camera hardware for capturing live video images, to be used for AR tracking and/or for overlaying virtual imagery onto the video images.
- *Display Hardware*: Either a handheld, head mounted, or projected display used to combine virtual images with images of the real world, creating the AR view.

---

[1] http://www.layar.com/

[2] http://www.wikitude.org/

[3] http://argon.junaio.com/

- *Networking*: Wireless or cellular networking support that will allow the mobile device to connect to remote data sources.
- *Sensor Hardware (optional)*: Additional GPS, compass or gyroscopic sensors that can be used to specify the user's position or orientation in the real world.

Using this technology and the associated software modules, the position and orientation of the user's viewpoint can be determined, and a virtual image created and overlaid on the user's view of the real world. As users change their viewpoint, the AR system updates their virtual world view accordingly. Thus, the basic AR process is:

1. Build a virtual world with a coordinate system identical to the real world.
2. Determine the position and orientation of the user's viewpoint.
3. Place the virtual graphics camera in that position and orientation.
4. Render an image of the physical world on the user's display.
5. Combine the virtual graphical overlay over the physical-world image.

1.2   Content of the Chapter

Although the hardware is readily available, there are a number of research and technical challenges that must be addressed before shared AR experiences are commonplace. In this chapter we provide an overview of MCAR systems with a particular focus on the history leading up the current systems, the typical technology used, and the important areas for future research. Later chapters in the book address specific topics in MCAR in more detail.

2   Mobile AR with Head Mounted Displays

The earliest mobile AR systems were based around head mounted displays rather than hand held mobile phones. Head Mounted Displays (HMDs) [14] were invented by Ivan Sutherland in the first AR system developed in 1965 [1]. Sutherland employed a physical optical system to combine the real world visual information with the virtual information. Currently the use of a digital camera to capture the visual information of the physical world allows the combination of both forms of visual information via the capabilities of modern graphics cards [15]. Using a HMD in conjunction with a head-position sensor and connected to a wearable computer, a user is able to see a large portable panoramic virtual information space surrounding them. A person can simply turn their head left, right, up, or down to reveal more information around them [16].

Figure 3 shows a conceptual image of a user within a wearable virtual information space, surrounded by pages of information. The combination of a head tracking sensor and HMD allows for the information to be presented in any direction from the user. However, a person's normal Field-of-View (FOV) is about 200

Fig. 3 A Wearable Virtual Information Space (from[47])

degrees [17] but typical commercial HMDs only have a FOV of between 30-60 degrees [18]. In spite of this, previous researchers such as Feiner and Shamash[19] and Reichlen [20] have demonstrated that a HMD linked to head movement can simulate a large "virtual" display.

A key feature of a wearable computer is the ability for a user to operate the computer while being mobile and free to move about the environment. When mobile, traditional desktop input devices such as keyboards and mice cannot be used, and so new user interfaces are required. Some currently available devices include: chord-based keyboards [21], forearm-mounted keyboards [22], track-ball and touch-pad mouse devices, gyroscopic and joystick-based mice, gesture detection of hand motions [23], vision tracking of hands [24], and voice recognition [25].

One particularly interesting control mechanism for navigating virtual information in a mobile AR interface is to use head movement. This should be intuitive, since it is how we normally explore the visual space around our bodies. The proprioceptive cues we get from muscles involved in head motion should aid navigation and object location. A head movement interface is a "direct" manipulation interface. AR use this concept by registering information in the physical world, so the user looking at a physical object can see overlaid graphical information.

The first demonstration of wearable AR system operating in an outdoor environment was the Touring Machine by Feiner et al. from Columbia University [26] (see Figure 4). This was based on a large backpack computer system with all the equipment attached to allow users to see virtual labels indicating the location of various buildings and features of the Columbia campus. Interaction with the system

Fig. 4 Touring Machine Hardware and User's View

was through the use of a GPS and a head compass to control the view of the world, and by gazing at objects of interest. Further interaction with the system was provided by a tablet computer with a web-based browser interface to provide extra information. As such, the system had all the key technology components mentioned in the previous section. In the next section we describe more of the history of mobile and handheld AR systems.

3 Mobile AR with Handheld Displays and Mobile Phones

In the previous section we described the core mobile AR technology and how the earliest prototype wearable AR system was developed using some of this technology. Now we give an expanded history of mobile AR systems, from backpack hardware to handheld devices.

Current MCAR systems have a rich history dating back to the mid-nineties, and Feiner's Touring Machine [26], described above. The Touring Machine was extended by Hollerer et al. for the placement of what they termed Situated Documentaries [27]. This system was able to show 3D building models overlaying the physical world, giving users the ability to see historical buildings that no longer existed on the Columbia University campus. Since that time other researchers explored the use of backpack and HMD based AR systems for outdoor gaming [28], navigation [29] and historical reconstructions [30], among other applications.

After several years of experimenting with backpack systems, handheld computers and personal digital assistants (PDA's) became powerful enough for mobile AR. Initially these were thin client applications, such as the AR-PDA project [31], in which the PDA was used to show AR content generated on a remote PC server and streamed wireless. Then in 2003 Wagner and Schmalstieg developed the first self contained PDA AR application [32], and The Invisible Train [33] was first handheld

Fig. 5 The Invisible Train

collaborative AR application (see Figure 5). Unlike the backpack systems, handheld AR interfaces are unencumbering and ideal for lightweight social interactions.

As AR applications began to appear on handheld devices, researchers also explored how to use mobile phones for Augmented Reality. Just like PDAs, the first mobile phones did not have enough processing power so researchers also explored thin client approaches with projects such as AR-Phone [34]. However, by 2004 phones were capable of simple image processing both Moehring [35] and Henrysson [10] developed marker based tracking libraries. This work enabled simple AR applications to be developed which ran entirely on the phone at 7-14 frames per second. Most recently, Wagner et al. [36] and Reitmayr et al. [37] have developed markerless tracking algorithms for mobile phone based AR systems (see Figure 6).

The Touring Machine and other wearable systems used GPS and inertial compass hardware to detect the user's position and orientation in the real world without relying on computer vision methods. The MARA project was the first that tried to provide the same functionality on a mobile phone [38]. An external sensor box was attached to the phone that contained a GPS and compass and bluetooth was used to wirelessly send the position and orientation data to the mobile phone (see Figure 7). This was then used to overlay virtual information over the live camera view of the phone. More recently a number of mobile systems have been developed that provide the same functionality, such as Layer[4] and Argon[5].

---

[4] http://www.layar.com/

[5] http://argon.gatech.edu/

Fig. 6 Mobile Phone Markerless AR Tracking

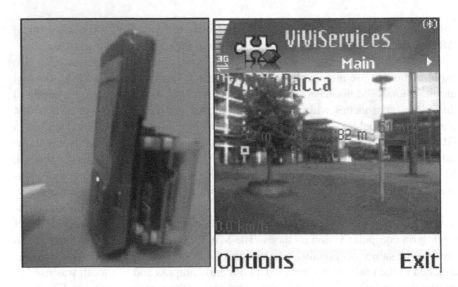

Fig. 7 MARA Hardware and Interface

4 Collaborative AR Systems

One of the most interesting uses for Augmented Reality is for enhancing face to face and remote collaboration. Current collaborative technology, such as video conferencing, often creates an artificial separation between the real world and shared

Fig. 8 Using the Shared Space System for Face to Face Collaborative AR

digital content, forcing the user to shift among a variety of spaces or modes of operation [39]. For example it is difficult with a desktop video conferencing system to share real documents or interact with on-screen 2D content while viewing the live video stream. Sellen summarizes several decades of telecommunications research by reporting that the main effect on communication is the presence of mediating technology rather than the type of technology used [40]. It is difficult for technology to provide remote participants with the same experience they would have if they were in a face to face meeting. However, Augmented Reality can blend the physical and virtual worlds and overcome this limitation.

At the same time as the development of early mobile AR systems, Schmalstieg et al. [41], Billinghurst et al. [42] and Rekimoto [43] were exploring early collaborative AR interfaces. Billinghurst et al.'s Shared Space work showed how AR can be used to seamlessly enhance face to face collaboration [44] (see Figure 8) and his AR Conferencing work showed how AR [42] could be used to create the illusion that a remote collaborator is actually present in a local workspace, building a stronger sense of presence than traditional video conferencing. Schmalstieg et al.'s Studierstube [41] software architecture was ideally suited for building collaborative and distributed AR applications. His team also developed a number of interesting prototypes of collaborative AR systems. Finally Rekimoto's Transvision system explored how a tethered handheld display could provide shared object viewing in an AR setting [43].

The first mobile AR collaborative system was the work of Hollerer [45] who added remote collaboration capabilities to the Touring Machine system, allowing a wearable AR user to collaborate with a remote user at a desktop computer.

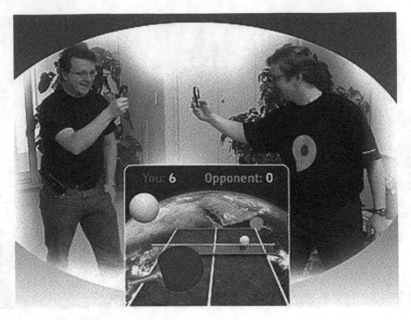

Fig. 9 Symball Application

Piekarski and Thomas [46] also added similar remote collaboration capabilities to their Tinmith system, once again between a wearable AR user and a colleague at a desktop computer. In contrast Reitmayr and Schmalstieg [13] developed an MCAR system that allowed multiple users with wearable AR systems to collaborate together in spontaneous ways, either face to face or in remote settings, using a backpack configuration. Billinghurst et al. developed a wearable AR conferencing space in which users could be surrounded by virtual images of people they are conferencing with and hear spatialized audio streams from their locations [47]. User studies found that the spatialized audio made it significantly easier to disambiguate multiple speakers and understand what they were saying. These projects showed that the same benefits that desktop AR interfaces provided for collaboration could also extend to the mobile platform.

Most recently, MCAR applications have been deployed on handheld systems and mobile phones. Wagner et al.'s Invisible Train [33] allowed several users to use PDAs in face to face collaboration and see virtual trains running on a real train track. They could collaborate to keep the trains running for as long as possible without colliding with each other. Hakkarainen and Woodward's "Symball" game [48] was a collaborative AR game in which each player could hit a virtual ball and play virtual table tennis with each other using a mobile phone. There was a virtual representation of the bat and ball superimposed over the real world (see Figure 9). Players could either play face to face, or remotely using internet connectivity, and a desktop player could also compete with a player on the mobile phone.

Fig. 10 Junaio AR View

5 Current MCAR Systems

As shown in the previous section there have been a number of research prototype MCAR systems that have developed. More recently a number of more sophisticated research and commercial systems have been created. In this section we describe several sample systems in more detail.

5.1 Junaio AR Browser

Since 2009 several companies have developed mobile phone AR browser applications. These use the GPS and compass sensors in smart phones to enable AR overlay on the live camera view. Unlike stand alone mobile AR experiences, AR Browser applications provide a generic browser interface and connect back to a remote server to load geo-located points of interest (POI) which are shown using virtual cues. Applications such as Layar allow users to subscribe to channels of interest (e.g. homes for sale) to show the set of POI that are most relevant.

Junaio[6] is a cross platform commercial AR browser that supports asynchronous collaboration, running on both iPhone and Android mobile phones. Like other AR browsers when users start it they can select a channel of interest and see virtual tags superimposed over the real world (see Figure 10). Users can see one of the following: an AR view, a list view of points of interest, or a map view where POIs are shown on a Google map. The AR view also shows a radar display showing where the POI is in relation to the user's position.

---

[6] http://www.junaio.com/

However, unlike most other AR browsers, Junaio also allows users to add their own content. Users are able to "Tag the World" where they can add 3D models, text notes, or 2D images at their current location. For example, a user could take a picture of a party they were at and then tag their location with the picture and text annotation. This picture and annotation is saved back to the Junaio server and can be seen by others who come to the same location. In this way Junaio supports mobile AR asynchronous collaboration. Virtual annotations can be made public so that anyone can see them, or private so they are only visible to the user's friends.

The main limitation with Junaio is that the interface for adding AR content is limited, so when a 3D model is created then the user cannot move, orient or scale it once it has been added to real world. The user is also limited to only using the pre-defined Junaio 3D models. However these are sufficient to add simple 3D tags to the real world, and the ability to take pictures and drop them in space is particularly useful for asynchronous collaboration.

5.2 The Hand of God

Modern command and control centers require support for temporally constrained collaborative efforts. We envision this technology to be intuitive and very straightforward to control. Picture in your mind a leader communicating to support people in the field, and they require a support person to walk to a particular position on a map. One straightforward method would be for the leader to point to a location on the map, and for a virtual representation to be shown to the field operative. This is an example of technology supporting through walls collaboration for the leader providing meaningful information to the operative in the field.

The Hand of God (HOG) system was constructed to connect indoor experts and operatives out in the field [49]. Figure 11 illustrates an indoor expert utilizing the HOG by pointing to places on a map. The indoor and outdoor users utilize a supplementary audio channel. An outdoor field worker makes use of a Tinmith wearable computer [50] (see Figure 2) and visualizes a 3D recreated virtual model of the indoor expert's hand geo-referenced at the indicated location on the map, as depicted in Figure 11. The indoor expert is able to rapidly and naturally communicate to the outdoor field operative, and give the outdoor user a visual waypoint to navigate to, see Figure 12. Physical props may be positioned on top of the HOG table, such as placing a signpost on a geo-referenced position (see Figure 13).

5.3 AR Tennis

The AR Tennis application was designed to support face to face collaboration on an AR game with mobile phones. In this case, users could sit across the table from one another and use their real mobile phones to view a virtual tennis court superimposed

Fig. 11 An indoor expert employing the Hand of God interface

Fig. 12 Head mounted display view seen by the outdoor user

over the real world between them [51]. Players could hit the ball to each other by moving their phone in front of the virtual ball (see Figure 14).

The application was run on Nokia N-95 phones using a Symbian port of the ARToolKit tracking library [10]. Players needed to place black square patterns on

Fig. 13 Physical props as signposts for the outdoor user

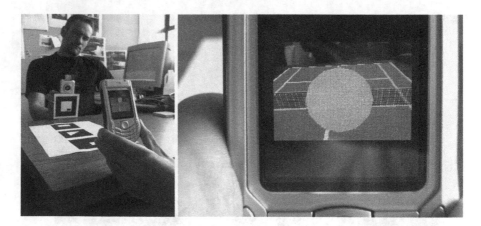

Fig. 14 AR Tennis

the table to support the AR tracking, and Bluetooth networking between the two phones was used to exchange game state information and ball position. A simple physics engine was integrated into the application to allow the ball to bounce realistically over the net. The game also supported multimodal feed-back. When the player's phone hit the virtual ball the sound of a ball being hit was played, and the phone vibrated to create the illusion that they were hitting a real ball.

The AR Tennis application was used to investigate how AR changed the face to face gaming experience [51]. A user study compared between people playing in an AR mode, in a graphics only mode where the user did not see video of the real world on their screen, and also in a non-face to face condition. Players were asked to collaborate to see how long they could keep a tennis volley going for. User's overwhelmingly preferred the face to face AR condition because they felt that they could more easily be aware of what the other player was doing and collaborate with them. They enjoyed being able to see the person they were playing with on the phone screen at the same time as the virtual court and ball.

6 Directions for Research

Although some MCAR systems have been developed, there is still a lot of research that must be conducted before such systems become commonplace. In particular there is important research that can be done in each of the following areas:

- Interaction Techniques
- Scaling Up to Large Numbers of Users
- Evaluation Methods
- New Devices
- New Design Methods

Investigating interaction techniques is a notable area of research for MCAR. Controlling the information in a mobile will require the creation of new user interfaces and input devices. Current technologies fall short on the requirements that these devices must be intuitive, non-intrusive, and robust. Many traditional input devices such as mice and keyboards are not suitable for mobile work outdoors, as they require a level flat surface to operate on.

The problem of registering virtual images with the user's view of the physical world is a main focus of current AR research. However, there is little previous work in the area of user interfaces for controlling AR systems in a mobile setting. Two major issues for the development of these user interfaces are as follows: firstly, registration errors will make it difficult for a user to point at or select small details in the augmented view, and secondly, pointing and selecting at a distance are known problems in VR and AR applications, compounded by the fact the user is outdoors with less than optimal tracking of their head and hands [52] [53].

Therefore, the investigation of new user interaction techniques is required. A key element of these new user interactions is that AR systems have a varying number of coordinate systems (physical world, augmented world, body relative, and screen relative) within which the user must work. Areas of investigation requires support for operations such as selecting small details in the augmentation, pointing/selecting at a distance, information overlays, text based messaging, and telepresence. While there have been some empirical user studies of existing commercial pointing devices for wearable computers: a handheld trackball, a wrist mounted touchpad, a handheld gyroscopic mouse and the Twiddler2 mouse [54], new input devices are required.

Current mobile phone technologies provide an interesting platform to support user interaction for mobile collaborative augmented reality. While the phone can support the entire AR technology requirements, it could also provide a convenient user input device for HMD style MCAR application. The phone has buttons and touch screen, but current phones have accelerometer and gyroscopic sensors in them. These sensors allow for the support of gestures. Depth cameras are becoming popular and also provide an opportunity to support hand gestures in a more complete fashion.

7   Conclusions

In this chapter we have described the history and development of mobile collaborative AR and set of examples from recent MCAR systems. As can be seen, MCAR systems have progressed rapidly from heavy back pack systems to mobile phones and handheld devices. From the MCAR research and commercial systems that have been developed there are a number of important lessons that can be learned that can inform the design of future mobile collaborative AR applications. For example, it is important to design around the limitations of the technology. Current mobile phones typically have noisy GPS and compass sensors, limited processing and graphics power, and a small screen. This means that the quality of the AR experience they can provide is very different from high end PC based systems. So successful MCAR systems do not rely on accurate tracking and complex graphics, but instead focus on how the AR cues can enhance the collaboration. For example in the AR Tennis application the graphics were very basic but the game was enjoyable because it encouraged collaboration between players.

One promising direction of future research is the concept of through walls collaboration [55] that enables users out in the field at the location where decisions have to be made to work in real time with experts indoors. The users out in the field have personal knowledge and context of the current issue, while the indoor experts have access to additional reference materials, a global picture, and more advanced technology. MCAR can supply a suitable hardware and software platform for these forms of systems.

References

1. Sutherland, I. *The Ultimate Display*. in *IFIP Congress*. 1965. New York, NY.
2. Fjeld, M. and B.M. Voegtli. *Augmented chemistry: An interactive educational workbench*. 2002: IEEE Computer Society.
3. Thomas, B., W. Piekarski, and B. Gunther. *Using Augmented Reality to Visualise Architecture Designs in an Outdoor Environment*. in *Design Computing on the Net* -http://www.arch.usyd. edu.au/kcdc/conferences/dcnet99. 1999. Sydney, NSW.
4. Cheok, A.D., et al., *Touch-Space: Mixed Reality Game Space Based on Ubiquitous, Tangible, and Social Computing*. Personal and Ubiquitous Computing, 2002. 6(5-6): p. 430-442.
5. Henderson, S.J. and S. Feiner. *Evaluating the benefits of augmented reality for task localization in maintenance of an armored personnel carrier turret*. in *8th IEEE International Symposium on Mixed and Augmented Reality*. 2009.

6. Billinghurst, M., K. Hirkazu, and I. Poupyrev. *The Magic Book - Moving Seamlessly between Reality and Virtuality*. in *IEEE Computer Graphics and Applications*. 2001.

7. Azuma, R.T., *Survey of Augmented Reality*. Presence: Teleoperators and Virtual Environments, 1997. 6.

8. Azuma, R., et al., *Recent Advances in Augmented Reality*. IEEE Computer Graphics and Applications, 2001. 21(6): p. 34-47.

9. Piekarski, W. and B.H. Thomas. Tinmith - *A Mobile Outdoor Augmented Reality Modelling System*. in *Symposium on Interactive 3D Graphics*. 2003. Monterey, Ca.

10. Henrysson, A. and M. Ollila, *UMAR: Ubiquitous Mobile Augmented Reality*, in *Proceedings of the 3rd international conference on Mobile and ubiquitous multimedia*. 2004, ACM: College Park, Maryland.

11. Wagner, D., et al. *Towards Massively Multi-User Augmented Reality on Handheld Devices*. in *Proc. of the Third International Conference on Pervasive Computing (Pervasive 2005)*. 2005. Munich, Germany.

12. Taehee, L. and H. Tobias, *Viewpoint stabilization for live collaborative video augmentations*, in *Proceedings of the 5th IEEE and ACM International Symposium on Mixed and Augmented Reality*. 2006, IEEE Computer Society.

13. Reitmayr, G. and D. Schmalstieg. *Mobile Collaborative Augmented Reality*. in *2nd ACM/IEEE International Symposium on Augmented Reality (ISAR'01)*. 2001. New York NY: IEEE.

14. Kiyokawa, K., et al. *An occlusion-capable optical see-through head mount display for supporting colocated collaboration*. 2003: IEEE Computer Society.

15. Rolland, J.P. and H. Fuchs, *Optical Versus Video See-Through Head-Mounted Displays in Medical Visualization*. Presence: Teleoperators and Virtual Environments, 2000. 9(3): p. 287-309.

16. Billinghurst, M., et al. *A wearable spatial conferencing space*. in *Second International Symposium on Wearable Computers*. 1998. Pittsburgh: IEEE.

17. Cutting, J.E. and P.M. Vishton, *Perceiving layout and knowing distances: The integration, relative potency, and contextual use of different information about depth*, in *Handbook of perception and cognition*, W. Epstein and S. Rogers, Editors. 1995, Academic Press: San Diego, Ca. p. 69-117.

18. Cakmakci, O. and J. Rolland, *Head-worn displays: a review*. Display Technology, Journal of, 2006. 2(3): p. 199-216.

19. Feiner, S. and A. Shamash. *Hybrid User Interfaces: Breeding Virtually Bigger Interfaces for Physically Smaller Computers*. in *Proceedings of the ACM Symposium on User Interface Software and Technology*. 1991.

20. Reichlen, B.A. *Sparcchair: A One Hundred Million Pixel Display*. in *Proceedings of IEEE Virtual Reality Annual International Symposium, VRAIS'93,*. 1993.

21. Richardson, R.M.M., et al. *Evaluation of conventional, serial, and chord keyboard options for mail encoding*. in *Human Factors Society*. 1987: Human Factors Society.

22. Thomas, B., S. Tyerman, and K. Grimmer. *Evaluation of Three Input Mechanisms for Wearable Computers*. in *1st Int'l Symposium on Wearable Computers*. 1997. Cambridge, Ma.

23. Fels, S. and G. Hinton, *Glove-TalkII: An Adaptive Gesture-to-Formant Interface*, in *CHI '97 Conference on Human Factors in Computing Systems: Mosaic of Creativity*. 1995: Denver, CO.

24. Takahashi, T. and F. Kishino, *Hand Gesture Coding Based on Experiments Using a Hand Gesture Interface Device*. ACM SIGCHI Bulletin, 1991. 23: p. 67-74.

25. Billinghurst, M., *Put that where? Voice and gesture at the graphics interface*. Computer Graphics (ACM), 1998. 32(4): p. 60-63.

26. Feiner, S., B. MacIntyre, and T. Hollerer. *A Touring Machine: Prototyping 3D Mobile Augmented Reality Systems for Exploring the Urban Environment*. in *1st Int'l Symposium on Wearable Computers*. 1997. Cambridge, Ma.

27. Hollerer, T., S. Feiner, and J. Pavlik. *Situated Documentaries: Embedding Multimedia Presentations in the Real World*. in *3rd Int'l Symposium on Wearable Computers*. 1999. San Francisco, Ca.

28. Thomas, B., et al. ARQuake: An Outdoor/Indoor Augmented Reality First Person Application. in IEEE 4th International Symposium on Wearable Computers. 2000. Atlanta, Ga.

29. Thomas, B., et al. A wearable computer system with augmented reality to support terrestrial navigation. in Digest of Papers. Second International Symposium on Wearable Computers. 1998. Los Alamitos, CA, USA.: IEEE Comput Soc.

30. Vlahakis, V., et al., Archeoguide: first results of an augmented reality, mobile computing system in cultural heritage sites, in Proceedings of the 2001 Conference on Virtual Reality, Archeology, and Cultural Heritage 2001 ACM Press: Glyfada, Greece p. 131-140

31. Geiger, C., et al. Mobile AR4ALL. in ISAR 2001, The Second IEEE and ACM International Symposium on Augmented Reality. 2001. New York.

32. Wagner, D. and D. Schmalstieg. First steps towards handheld augmented reality. in Proceedings. Seventh IEEE International Symposium on Wearable Computers. 2003.

33. Wagner, D., T. Pintaric, and D. Schmalstieg, The invisible train: a collaborative handheld augmented reality demonstrator, in ACM SIGGRAPH 2004 Emerging Technologies. 2004, ACM: Los Angeles, California.

34. Cutting, D.J.C.D., M. Assad, and A. Hudson. AR phone: Accessible augmented reality in the intelligent environment. in OZCHI 2003. 2003. Brisbane.

35. Moehring, M., C. Lessig, and O. Bimber. Video SeeThrough AR on Consumer Cell Phones. in International Symposium on Augmented and Mixed Reality (ISMAR'04). 2004.

36. Wagner, D., et al. Pose tracking from natural features on mobile phones. in Proceedings of the 7th IEEE/ACM International Symposium on Mixed and Augmented Reality. 2008: IEEE Computer Society.

37. Reitmayr, G. and T.W. Drummond. Going out: Robust modelbased tracking for outdoor augmented reality. 2006.

38. Kähäri, M. and D.J. Murphy. Mara: Sensor based augmented reality system for mobile imaging device. in 5th IEEE and ACM International Symposium on Mixed and Augmented Reality. 2006. Santa Barbara.

39. Ishii, H., M. Kobayashi, and K. Arita, Iterative Design of Seamless Collaboration Media. Communications of the ACM, 1994. 37(8): p. 83-97.

40. Sellen, A., Remote Conversations: The effects of mediating talk with technology. Human Computer Interaction, 1995. 10(4): p. 401-444.

41. Schmalstieg, D., et al. Studierstubean environment for collaboration in augmented reality. in Collaborative Virtual Environments 1996 (CVE'96). 1996. Nottingham, UK.

42. Billinghurst, M., S. Weghorst, and T. Furness. Shared Space: Collaborative Augmented Reality. in Workshop on Collaborative Virtual Environments 1996 (CVE 96). 1996 Nottingham, UK.

43. Rekimoto, J. TransVision: A Hand-Held Augmented Reality System for Collaborative Design. in Virtual Systems and Multimedia. 1996.

44. Billinghurst, M., et al. Mixing realities in shared space: An augmented reality interface for collaborative computing. in ICME 2000. 2000.

45. Hollerer, T., et al., Exploring MARS: Developing Indoor and Outdoor User Interfaces to a Mobile Augmented Reality System. Computers and Graphics, 1999. 23(6): p. 779-785.

46. Piekarski, W. and B.H. Thomas. Tinmith-Hand: unified user interface technology for mobile outdooraugmented reality and indoor virtual reality. in IEEE Virtual Reality, 2002. 2002. Orlando, FL: IEEE.

47. Billinghurst, M., J. Bowskill, and J. Morphett, Wearable communication space. British Telecommunications Engineering, 1998. 16(pt 4): p. 311-317.

48. Hakkarainen, M. and C. Woodward. SymBall - Camera driven table tennis for mobile phones. in ACM SIGCHI International Conference on Advances in Computer Entertainment Technology (ACE 2005). 2005. Valencia, Spain.

49. Stafford, A., W. Piekarski, and B.H. Thomas. Implementation of God-like Interaction Techniques For Supporting Collaboration Between Indoor and Outdoor Users in 5th IEEE and ACM International Symposium on Mixed and Augmented Reality. 2006. Santa Barbara.

50. Piekarski, W., et al. Tinmith-Endeavour - A Platform For Outdoor Augmented Reality Research. in 6th Int'l Symposium on Wearable Computers. 2002. Seattle, Wa.

51. Henrysson, A., M. Billinghurst, and M. Ollila, AR Tennis, in ACM SIGGRAPH 2006 Sketches. 2006, ACM: Boston, Massachusetts.

52. Thuong, N.H. and H.T. Bruce, In-situ refinement techniques for outdoor geo-referenced models using mobile AR, in Proceedings of the 2009 8th IEEE International Symposium on Mixed and Augmented Reality. 2009, IEEE Computer Society.

53. Jason, W., D. Stephen, and H. Tobias, Using aerial photographs for improved mobile AR annotation, in Proceedings of the 5th IEEE and ACM International Symposium on Mixed and Augmented Reality. 2006, IEEE Computer Society.

54. Handykey Corporation, The Twiddler. 1996: 141 Mt. Sinai Avenue, Mt. Sinai, New York 11766, USA.

55. Thomas, B.H. and W. Piekarski, Through Walls Collaboration. IEEE Pervasive Computing, 2009. 9(3): p. 42-49.

Current Status of Standards for Augmented Reality

Christine Perey, Timo Engelke, and Carl Reed

Abstract This chapter discusses the current state, issues, and direction of the development and use of international standards for use in Augmented Reality (AR) applications and services. More specifically, the paper focuses on AR and mobile devices. Enterprise AR applications are not discussed in this chapter. There are many existing international standards that can be used in AR applications but there may not be defined best practices or profiles of those standards that effectively meet AR development requirements. This chapter provides information on a number of standards that can be used for AR applications but may need further international agreements on best practice use.

1 Introduction

Standards frequently provide a platform for development; they ease smooth operation of an ecosystem in which different segments contribute to and benefit from the success of the whole, and hopefully provide for a robust, economically-viable, value chain. One of the consequences of widespread adoption of standards is a baseline of interoperability between manufacturers and content publishers. Another is the ease of development of client applications.

In most markets, standards emerge during or following the establishment of an ecosystem, once a sufficient number of organizations see market and business value in interoperating with the solutions or services of others.

Publishers of content that support AR applications are motivated to make their content available when there is an assortment of devices that support the content for different use cases and this translates into the maximum audience size. Standards

C. Perey (✉)
Spime Wrangler, PEREY Research & Consulting
e-mail: cperey@perey.com

L. Alem and W. Huang (eds.), *Recent Trends of Mobile Collaborative Augmented Reality Systems*, DOI 10.1007/978-1-4419-9845-3_2,
© Springer Science+Business Media, LLC 2011

enable such device and use case independence, thereby reducing implementation costs and mitigating investment risks.

As of early 2011, the mobile AR solutions available to users and developers are based on a mixture of proprietary and open standards protocols and content encodings, without interoperable – or standards based - content, platforms or viewing applications. It is a field of technology silos and, consequently fragmented markets.

For mobile collaborative AR, the needs for standards are compounded by the fact that the content of shared interest must travel over a communications "bridge" which is, itself, established between end points between and through servers, client devices and across networks. The more interoperable the components of the end-to-end system are, the less the need for the participants in a collaborative session to use technologies provided by the same manufacturer. More interoperability translates directly into more enabled people, hence more potential collaborators, and more service and application providers.

2   Guiding principles of an open AR industry

Open AR, or interoperable systems for viewing content in real time in context, is a design goal for the evolution of the AR market. Currently, there are numerous standards that can be used in the development and deployment of open AR applications and services. However, there are still interoperability gaps in the AR value chain. Further, work needs to be done to determine best practices for using existing international standards. In some cases in which there are interoperability gaps, new standards will need to be defined, documented, and tested. Developing new standards and pushing them through the development process required in a standards development organization may not be appropriate for the needs of the AR community. In this case, perhaps profiles of existing standards would be more appropriate. Further, the development of an over-arching framework of standards required for AR may be beyond the resources of any single body. And, as AR requires the convergence of so many technologies, there are numerous interoperability challenges. As such, there will not be one "global" AR standard. Instead, there will be a suite of standards for use in AR applications.

Many technology participants in the AR ecosystem desire to leverage existing standards that solve different interoperability issues. For example, standards which permit an application to learn the locations of users, how to display objects on the users' screen, how to time stamp every frame of a video, how to use the users' inputs for managing behaviors, and which are proven and optimized... to be extended to address new or related issues which AR raises.

One of the strongest motivators for a cross-standard, multi-consortium and open discussion about standards and AR is time-to-market. Re-purposing existing content and applications is critical. The use of existing standards or profiles of these standards is driven by the need to avoid making mistakes and also use of currently deployed and proven (and emerging) technologies to solve/address urgent issues for AR publishers, developers and users.

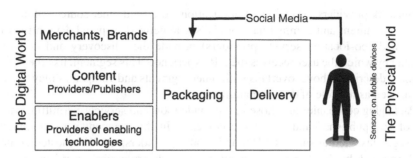

Fig. 1 Ecosystem of mobile AR Segments

However, the time-to-market argument is only valid if one assumes that there is a motivation/agreement on the part of most or all members of the ecosystem that having open AR—the opposite of technology "silos"—is a good thing. Based on the participation of academic and institutional researchers, companies of all sizes and industry consortia representing different technology groups, there is agreement across many parts of the AR ecosystem regarding the need for standards[1].

2.1   Suggested model of a General AR ecosystem

The AR ecosystem is composed of at least six interlocking and interdependent groups of technologies. Figure 1 shows how these interlocking and interdependent groups bridge the space between the digital and physical worlds in a block diagram.

Beginning on the far right side of the figure, there is the "client" in the networked end-to-end system. The user holds or wears the client, a device that provides (1) an interface for the user to interact with one or more services or applications and the information in the digital world, and (2) integrates an array of real-time sensors. The sensors in the users' devices detect conditions in the users' environments as well as in some cases the users' inputs. There may also be sensors (e.g., cameras, pressure sensors, microphones) in the environment to which the applications could provide access. The client device is also the output for the user, permitting visualization or other forms of augmentation such as sounds or haptic feedback.

Manufacturers of components and finished AR-capable devices (e.g., Smartphones) occupy both the client segment of the ecosystem as well as, in some cases, the "technology enablers" segment to the far left of the figure. At the right side, the client devices are frequently tightly connected to the networks.

---

[1]This conclusion is based on the results of discussions at two recent multi-participant AR workshops.

Network providers, providers of application stores and other sources of content (e.g., government and commercial portals, social networks, spatial data infrastructures, and geo-location service providers) provide the "discovery and delivery" channel by which the user receives the AR experience. This segment, like the device segment described above, overlaps with other segments and companies may occupy this as well as the role of device manufacturer.

Packaging companies are those that provide tools and services permitting viewing of any published and accessible content. In this segment we can imagine sub-segments such as the AR SDK and toolkit providers, the Web-hosted content platforms and the developers of content that provide professional services to agencies, brands and merchants.

Packaging companies provide their technologies and services to organizations with content that is suitable for context-driven visualization. For the case of collaborative AR, this is probably not an important segment since, in effect, users themselves are the creators of content[2].

For some purposes, packaging companies rely on the providers of enabling technologies, the segment represented in the lower left corner of the figure. Like the packaging segment, there are sub-segments of enabling technologies (e.g., semiconductors, sensor providers, algorithms, etc). This segment is rich with existing standards that can and are already being assimilated by the companies in the packaging segment of the ecosystem.

Content providers include a range of public, proprietary (commercial), and user-provided data. Traditionally, proprietary content was the primary source of content for use in AR applications. More recently, more and more content is being provided by government agencies (e.g. traffic data or base map information) and volunteered sources (e.g. Open Street Map). An excellent example of this evolution from proprietary to a mixed content platform is the map data used in AR applications.

The traditional AR content providers, brands and merchants who seek to provide their digital information to users of AR-enabled devices, are reluctant to enter the AR ecosystem until they feel that the technologies are stable and robust. The adoption of standards for content encoding and access of content by platforms and "packaging segment" providers is a clear indicator of a certain market maturity for which content providers are waiting. The use of standard interfaces and encodings also allows application providers to access content from many more sources, including proprietary, user-provided, and government sources.

Furthermore, content providers and end users of AR applications will benefit when AR content standards are available to express the provenance and quality of the source content in a consistent fashion.

---

[2] Also known as "user provided" content.

3 AR Requirements and Use Cases

For the proper development of standards for AR, there needs to be a very clear understanding of AR requirements and use cases. Different domains have different AR requirements. For example, mass market mobile AR tourist applications requirements and related use cases may be different from those required by first responders in an emergency scenario. Such analysis will also permit identification of the most common requirements. By specifying use cases and requirements, standards organizations will have the information necessary to determine which standards can best be used in given situations or workflows. Given that different AR ecosystem segments have different requirements for standards, different standards bodies and industry consortia have been working on various aspects of the AR standards stack. Therefore, stronger collaboration between the various standards bodies is required. By conducting face-to-face open meetings of interested parties, such as AR DevCamps and the International AR Standards Meetings, people from vastly different backgrounds are convening to exchange (share) information about what they have seen succeed in their fields and how these may be applied to the challenges facing interoperable and open AR. Having a common set of use cases and related requirements provides the "lingua-franca" for collaboration and discussion.

This collaboration between Standards Development Organizations (SDOs) and their related expert communities is crucial at this juncture in the growth of AR. If we can benefit from the experience of those who wrote, who have implemented and who have optimized a variety of today's most popular standards already in use for AR or AR-like applications (OpenGL ES, JSON, HTML5, KML, GML, CityGML, X3D, etc.), the goal of interoperable AR will be more quickly achieved and may avoid costly errors.

Discussions on the topic of standards to date indicate that the development of standards specifically for AR applications is necessary in only a small number of cases. Instead, re-purposing (profiles) and better understanding of existing standards from such organizations as the Khronos Group (Khronos, 2011), the Web3D Consortium, the World Wide Web Consortium (W3C), the Open Geospatial Consortium (OGC), the 3GPP, the Open Mobile Alliance (OMA) and others is the way to proceed with the greatest impact and assurance that we will have large markets and stable systems in the future.

3.1 AR Content publisher requirements

The content sources from which the future AR content will be produced and deployed are extremely varied. They range from multi-national information, news and media conglomerates, device manufacturers, to national, state, and local government organizations, user-provided content, to individual content developers who wish to share their personal trivia or experiences. Clearly, publisher sub-segments will have needs for their specialized markets or use cases.

As a broad category, the content publisher's needs are to:

- Reach the maximum potential audience with the same content,
- Provide content in formats which are suited to special use cases (also known as re-purposing),
- Provide accurate, up-to-date content,
- Control access to and limit uncontrolled proliferation (pirating) of content.

For content publishers, a simple, lightweight markup language that is easily integrated with existing content management systems and offers a large community of developers for customization, is highly desirable.

In collaborative AR, the case can be made that the users themselves are the content that is being enhanced. In this view, the users will rely on real time algorithms that convert gestures, facial expressions, and spoken and written language into objects or content, which is viewed by others at a distance.

Real time representation of 2D and 3D spaces and objects at a distance will rely on projection systems of many types and for remote commands to appear in the view of local users. These remote commands could leverage the existing work of the multimedia telecommunications manufacturers and videoconferencing systems adhering to the ITU H.3XX standard protocols.

3.2 Packaging segment requirements

This is the segment of the AR ecosystem in which the proprietary technology silos are most evident at the time of this study and where control of the content development platforms is highly competitive. There are the needs for differentiation of the providers of tools and platforms that are substantially different than those of the professional service providers who use the tools to gain their livelihoods.

Tools and platform providers seek to be able to:

- Access and process content from multiple distributed repositories and sensor networks. This may include repackaging for efficiency. However, for certain content types, such as maps or location content, the ability to access the content closest to source allows the end user to use the latest, best quality content;
- Offer their tools and platforms to a large (preferably existing) community of developers who develop commercial solutions for customers;
- Integrate and fuse real time sensor observations into the AR application;
- Quickly develop and bring to market new, innovative features that make their system more desirable than a competitor's or a free solution.

Professional developers of content (the service providers who utilize the SDKs and platforms for publishing) seek to be able to:

- Repurpose existing tools and content (managing costs as well as learning curves) to just make an "AR version" of their work

- Provide end users rich experiences that leverage the capabilities of an AR platform but at the same time have features tying them to the existing platforms for social networking, communications, navigation, content administration and billing.

3.3 AR system and content users

This is the most diverse segment in the AR stack in the sense that users include all people, related services, and organizations in all future scenarios. It is natural that the users of AR systems and content want to have experiences leveraging the latest technologies and the best, most up-to-date content without losing any of the benefits to which they have grown accustomed.

In the case of collaborative mobile AR users, they seek to:

- Connect with peers or subject matter experts anywhere in the world over broadband IP networks,
- Show and manipulate physical and local as well as virtual objects as they would if the collaborator were in the same room, and
- Perform tasks and achieve objectives that are not possible when collaborators are in the same room.

4 Approaches to the AR Standards challenge

To meet the needs of developers, content publishers, platform and tool providers and users of the AR ecosystem, the experts in hardware accelerated graphics, imaging and compute, cloud computing and Web services, digital data formats, navigation, sensors and geospatial content and services management and hardware and software devices must collaborate.

4.1 Basic tools of the standards trade

Standards that are or will be useful to the AR ecosystem segments will leverage know-how that is gained through both experimentation, and creation of concrete open source, commercial and pre-commercial implementations. In most standards activities, the process of developing a recommendation for standardization begins with development of core requirements and use cases. This work is then followed by development of a vocabulary (terms and definitions), information models, abstract architectures and agreement on the principle objectives.

An AR standards gap analysis must be performed. The results of the gap analysis combined with known requirements will reveal where the community should concentrate future standardization efforts. Finally, any new standards work designed to fill the gaps can begin in existing standards organizations to support AR experiences.

4.2   Standards gap analysis

A gap analysis begins with detailed examinations of available standards and to determine which standards are close to and which are distant from meeting the AR ecosystem requirements. The gap analysis process began during the International AR Standards Meeting in Seoul, October 11-12, 2010.

The gap analysis exercise divided the scope of the problem into two large spaces: those related to content and software, and those that are most relevant to hardware and networks.

Existing standards search results were grouped according to whether the standard addresses a content/software service related issue or a network and hardware issue. In some cases, there is overlap.

5   Content-related standards

5.1   Declarative and imperative approaches to Content Encoding

First, it is important to the success of the gap analysis to clarify the differences between the declarative and the imperative approaches of standards.

The imperative approach of description usually defines how something is to be computed, like code. It features storage, variables, states, instructions and flow control. Usually it benefits from a high potential of possibilities and user driven variations. In use, imperative code can be designed in any manner, as long as they conform to the common rules of the interpreting background system. A typical example is JavaScript (JS), a highly popular implementation of the ECMAScript (ECMA-262) language standard, which is part of every Web browser on mobile and desktop systems today.

Declarative approaches are more restrictive and their design usually follows a strict behavior scheme and structure. They consist of implicit operational semantics that are transparent in their references. They describe what is to be computed. Declarative approaches usually do not deliver states and, thus, dynamic systems are more difficult to achieve using declarative approaches. On the other hand, they tend to be more transparent and easy to use and generate. A common declarative language in use today is the W3C XML standard which defines a hierarchical presentation of elements and attributes. Another coding form for declarative data is the JavaScript Object Notation (JSON), which benefits from being a lightweight

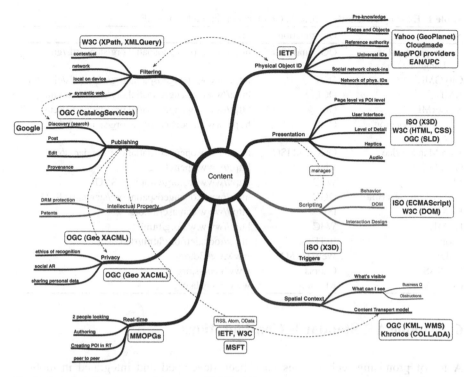

Fig. 2 Standards Landscape of Impact to Mobile AR

and easy-to-port data interchange format. It builds upon the ECMAScript specification.

Both approaches, or a combination of these, are likely to be used for creation of AR content encoding and payload standards.

5.2 Existing Standards

There are many content or payload encoding related standards that could be used for AR applications. The diagram below suggests an initial inventory of such standards and their possible relationships.

The table below shows Geo Information System (GIS)-based standards and other standards used within the system along with the Web addresses where further definitions can be found. The organizations mentioned are potential providers of experience and knowledge in specialized fields. Today's mapping software is usually based on these standards; consequently, AR services and applications that rely on the user's location also leverage these standards. There are also many standards that define the position and interactivity of virtual objects in a user's visual space.

Table 1 Existing Standards for Use in Geo-location-based Mobile AR

Standards	Organization	url
Geography Markup Language	OGC and ISO	http://www.opengeospatial.org/standards/gml
CityGML	OGC	http://www.opengeospatial.org/standards/citygml
KML	OGC	http://www.opengeospatial.org/standards/kml
SensorML	OGC	http://www.opengeospatial.org/standards/sensorml
Sensor Observation Service	OGC	http://www.opengeospatial.org/standards/SOS
Web Map Service	OGC and ISO	http://www.opengeospatial.org/standards/wms
OpenGL	Khronos	http://www.opengl.org/
SVG	W3C	http://www.w3.org/Graphics/SVG/
Style Layer Descriptor	OGC	http://www.opengeospatial.org/standards/SLD
ECMAScript	ISO	http://www.ecmascript.org/
HTML	W3C	http://www.w3.org/html/
Atom	IETF	http://tools.ietf.org/html/rfc4287
X3D	Web3D/ISO	www.web3d.org
GeoRSS	Georss	www.georss.org
COLLADA	Khronos	www.collada.org

6 Mobile AR Standards Considerations

A lot of promising technologies have been developed and integrated in mobile device hardware. New sensor technologies allow delivering sensor data for mobile AR applications in a format that can be processed. The processing power in mobile devices, network and memory bandwidth supporting the latest mobile devices and applications have expanded exponentially in recent years. Software frameworks and platforms for mobile application development have also made huge advances, permitting developers to create new user experiences very quickly. This provides a huge potential for context- and location-aware AR applications or applications that are extended to take advantage of new capabilities.

Although these developments have accelerated the growth of the number and diversity of mobile AR applications, this growth has come at a cost. There is clearly a lack of standards for implementing mobile AR applications for users of multiple, different platforms and in different use scenarios.

In the next subsections we describe the use of standards in mobile AR sensing, processing and data presentation.

6.1 Mobile AR and Sensors

A sensor is an entity that provides information about an observed property as its output. A sensor uses a combination of physical, chemical or biological means in order to estimate the underlying observed property. An observed property is an identifier or description of the phenomenon for which the sensor observation result provides an estimate of its value. Satellites, cameras, seismic monitors, water temperature and flow monitors, accelerometers are all examples of sensors. Sensors may be in-situ, such as an air pollution monitoring station, or they may be dynamic, such as an unmanned aerial vehicle carrying a camera. The sensor observes a property at a specific point in time at a specific location, i.e. within a temporal and spatial context. Further, the location of the sensor might be different from the location of the observed property. This is the case for all remote-observing sensors, e.g. cameras, radar, etc.

From a mobile AR perspective, sensors may be onboard (in the device) or external to the device and accessed by the AR application as required, Regardless, all sensors have descriptions of the processes by which observations and measurements are generated, and other related metadata such as quality, time of last calibration, and time of measurement. The metadata, or characteristics, of the sensor are critical for developers and applications that require the use of sensor observations. The ability to have a standard description language for describing a sensor, its metadata, and processes will allow for greater flexibility and ease of implementation in terms of accessing and using sensor observations in AR applications.

Sensors behave differently on different and distinct device types and platforms. Due to differences in manufacturing tolerances or measurement processes, dynamic, or mobile, sensor observations may also be inconsistent even when observing the same phenomenon. Calculation of user location indoors is one example where wide variability may occur. Different location measurement technologies provide different levels of accuracy and quality. The problem is exacerbated by a variety of factors, such as interference from other devices, materials in the building, and so forth.

Approaches combining inaccurate geo-positioning data along with computer vision algorithms are promising for increasing accuracy of mobile AR, but require the definition of new models for recognition, sensor-fusion and reconstruction of the pose to be defined. Ideally, an abstraction layer which defines these different "sensor services" with a well-defined format and its sensor characteristics, would address existing performance limitations.

For some AR applications, real-time processing of vision-based data is crucial. In these cases, direct camera data is not appropriate for processing in a high level programming environment, and should be processed on a lower level. Since the processing is performed at the lower level, algorithms that create an abstracted sensor data layer for pose will be beneficial. In summary, sensor fusion and interpretation can happen on different levels of implementation[3].

Sensors with different processing needs can contribute to the final application outcome. In parallel, the higher level application logic may benefit from taking data from multiple sensors into account. Standards may provide direct access to sensor data or higher-level semantic abstractions of it.

---

[3] As an example of fusion requirements, consider the OGC "Fusion Standards Study Engineering Report". http://portal.opengeospatial.org/files/?artifact_id=36177

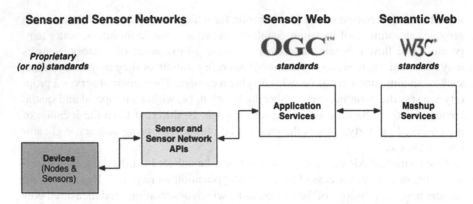

Fig. 3 The OGC Sensor Web Enablement Standards Landscape

A critical content source for many AR applications, independent of domain, will be near real-time observations obtained from in-situ and dynamic sensors. Examples of in-situ sensors are traffic, weather, fixed video, and stream gauges. Dynamic sensors include unmanned aerial vehicles, the mobile human, and satellites. Already, the vast majority of content used in AR applications is obtained via some sensor technology, such as LIDAR[4], that is subsequently processed and stored in a content management system. There are many other sources of sensor data that are (or will be) available on demand or by subscription. These sensor observations need to be fused into the AR environment in real time as well. As such, there is a need for standards that enable the description, discovery, access, and tasking of sensors within the collaborative AR environment.

The following is a simple diagram depicting one widely implemented sensor standards landscape.

A sensor network is a computer-accessible network of many, spatially-distributed devices using sensors to monitor conditions at different locations, such as temperature, sound, vibration, pressure, motion or pollutants. A Sensor Web refers to web-accessible sensor networks and archived sensor data that can be discovered and accessed using standard protocols and APIs.

There is a suite of standards that support Sensor Web Enablement (SWE) maintained by the OGC. SWE standards include:

1. Observations & Measurements Schema (O&M) – Standard models and XML Schema for encoding observations and measurements from a sensor, both archived and real-time.
2. Sensor Model Language (SensorML) – Standard models and XML Schema for describing sensors systems and processes; provides information needed for discovery of sensors, location of sensor observations, processing of low-level sensor observations, and listing of taskable properties.

---

[4]LIDAR is an acronym for LIght Detection And Ranging.

3. Transducer Markup Language (TransducerML or TML) – The conceptual model and XML Schema for describing transducers and supporting real-time streaming of data to and from sensor systems.
4. Sensor Observations Service (SOS) - Standard web service interface for requesting, filtering, and retrieving observations and sensor system information. This is the intermediary between a client and an observation repository or near real-time sensor channel.
5. Sensor Planning Service (SPS) – Standard web service interface for requesting user-driven acquisitions and observations. This is the intermediary between a client and a sensor collection management environment.

These standards could be used for low level descriptions of sensors and their fusion, and combined with visual processing for pose estimation and tracking, supply automatically-generated data for augmenting the users' immediate environment.

6.2 Mobile AR processing standards

Considerable standards work has previously been done in the domains of situational awareness, sensor fusion, and service chaining (workflows). This work and some of these standards can be applied to processes in an AR workflow. For example, the OGC Web Processing Service (WPS) provides rules for standardizing how inputs and outputs (requests and responses) for geospatial processing services, such as polygon overlay. The standard also defines how a client can request the execution of a process, and how the output from the process is handled. It defines an interface that facilitates the publishing of geospatial processes and clients' discovery of and binding to those processes. The data required by the WPS can be delivered across a network or they can be available at the server. The WPS can be used to "wrap" processing and modeling applications with a standard interface. WPS can also be used to enable the implementation of processing workflows.

In addition, the OpenGIS Tracking Service Interface Standard supports a very simple functionality allowing a collection of movable objects to be tracked as they move and change orientation. The standard addresses the absolute minimum in functionality in order to address the need for a simple, robust, and easy-to-implement open standard for geospatial tracking.

Other approaches for descriptions of vision-based tracking environments with its visual, camera constraints have been made first through Pustka et al., by introducing spatial relationship patterns for augmented reality environment descriptions.

There are several standards that could be applied to the presentation and visualization workflow stack for AR applications. There are service interfaces that an AR application can use to access content, such as a map for a specific area. Then there are lower level standards that enable standard mechanisms for rendering the content on the device.

Not all AR requires use of 3D. In some use cases and, especially on low processor devices unable to render 3D objects, 2D annotations are preferable. A very convenient declarative standard for the description of 2D annotation is W3C's

Scalable Vector Graphic (SVG) standard or even HTML. This is a large field in which many existing standards are suitable for AR use.

6.3   Mobile AR Acceleration and Presentation Standards

Augmented Reality is highly demanding in terms of computation and graphics performance. Enabling truly compelling AR on mobile devices requires efficient and innovative use of the advanced compute and graphics capabilities becoming available in today's smartphones.

Many mobile AR applications make direct and/or indirect use of hardware for acceleration of computationally complex tasks and, since the hardware available to the applications varies from device to device, standard Application Programming Interfaces (APIs) reduce the need for customization of software to specific hardware platforms.

The Khronos Group is an industry standards body that is dedicated to defining open APIs to enable software to access high-performance silicon for graphics, imaging and computation. A typical AR system with 3D graphics uses several Khronos standards and some that are under development. For example:

- OpenGL ES is a streamlined version of the widely respected desktop OpenGL open standard for 3D graphics. OpenGL ES is now being used to provide advanced graphics on almost every 3D-capable embedded and mobile device;
- OpenMAX provides advanced camera control, image and video processing and flexible video playback capabilities;
- OpenCL provides a framework for programming heterogeneous parallel CPU, GPU and DSP computing resources. Already available in desktop machines, OpenCL is expected to start shipping on mobile devices in 2012, becoming mainstream in mobile in 2013;
- OpenCV is a widely used imaging library that will potentially join Khronos to define an API to enable acceleration of advanced imaging and tracking software;
- StreamInput is a recently initiated Khronos working group that is defining a high-level, yet flexible framework for dealing with multiple, diverse sensors, enabling system-wide time-stamping of all sensor samples and display outputs for accurate sensor synchronization, and presenting high-level semantic sensor input to applications;
- COLLADA is an XML-based 3D asset format that can contain all aspects of 3D objects and scenes including geometry, textures, surface effects, physics and complex animations. COLLADA can be used to transmit 3D data over a network – or can be encoded to suit a particular application or use case;
- EGL is a window and surface management API that acts as an interoperability hub between the other Khronos APIs – enabling images, video and 3D graphics to be flexibly and efficiently transferred for processing and composition;
- OpenSL ES, not shown on the visual flow diagram, is an advanced native audio API that provides capabilities from simple alert sounds, through high-quality audio mixing through to full 3D positional audio that interoperates well with OpenGL ES 3D visuals.

Using these Khronos APIs, it is now becoming possible to create a mobile AR application that uses advanced camera and sensor processing from any device using the APIs to feed an accelerated image processing pipeline, that in-turn inputs to an accelerated visual tracker, that drives an advanced 3D engine that flexibly composites complex 3D augmentations into the video stream – all accompanied with a fully synchronized 3D audio stage.

6.4 Making Browser's AR Capable

Many developers and middleware vendors have strived to create application frameworks to enable content that is portable across diverse hardware platforms. The collection of standards and initiatives known as HTML5 is turning the browser into an application platform capable of accessing platform resources such as memory, threads and sensors – creating the opportunity for web content to be highly capable as well as widely portable.

Enabling browsers to support AR requires the Khronos native system resources be made available to web developers, typically leveraging the main components of a modern browser: JavaScript, the DOM and CSS.

The first 'connect point' between the native world of C-based acceleration APIs and the web is WebGL from the Khronos Group that defines a JavaScript binding to the OpenGL ES graphics rendering API – providing web developers the flexibility to generate any 3D content within the HTML stack without the need for a plug-in.

6.5 Declarative Programming

Some content creators, particularly those working on the web, prefer to use declarative abstractions of 3D visualizations. For instance, X3D is an ISO standard that allows the direct description of scenes and flow graphs. X3D can be used in conjunction with other standards, such as MPEG-4 and OGC CityGML. Besides a scene-graph description of objects, X3D also allows logic to be described in a declarative way.

Another declarative approach for describing soundscapes is the Audio Markup Language (A2ML) proposed by Lemordant.

7 Mobile AR architecture options

In the creation of AR applications, many disciplines may converge: computer vision recognition and tracking, geospatial, 3D etc. Algorithms for computer vision and sensor fusion are provided by researchers and software engineers or as third party services, while content for presentation and application logic are likely to be created by experience designers.

The most common standard for low-level application development on mobile devices today is C++. The problem with developing in C++ is that presentation and sensor access is different on each device and even the language used to access different sensors is not standardized. This results in a huge effort on the part of application developers for maintaining code bases. Thus, it would be desirable to have declarative descriptions of how AR logic is processed in order to have reusable blocks for AR application development. It would also be desirable to have a common language for definition of AR content.

All smart phones and other mobile devices today have a Web browser. Many already support elements of the HTML5 standard and this trend towards full HTML 5 support will continue. This represents a very clear and simple option for use by AR developers. Within a Web browser, JavaScript directly allows accessing the Document Object Model (DOM) and thus observing, creating, manipulating and removing of declarative elements. An option in standardization for AR could be the integration of declarative standards and data directly into the DOM.

A complete X3D renderer that builds on the WebGL standard and uses JavaScript has been implemented; it is called X3Dom (WebGL is a JavaScript interface for encapsulating OpenGL ES 2). X3Dom is completely independent of the mobile platform on which the application will run. This does not imply that an output solution, such as X3Dom, is all that's needed to make a universal viewer for AR. For example, a convenient interface extension for distributed access of real-time processed, concrete or distributed sensor data would be required.

A promising way for data synchronization of collaborative AR data within the network will be using existing standards, like the Extensible and Presence Protocol (XMPP), which has been established by the IETF. It has been implemented in chat applications and already delivers protocols for decentralized discovery, registration and resource binding. The work of the ARWAVE project could produce interesting results for mobile collaborative AR in the future.

8   Future mobile collaborative AR standards architectures

While in most of AR applications content is simply rendered over a user's camera view, AR can also be much more complex, particularly for mobile collaborative scenarios.

Technologies can, in the future, produce "reality filtering" to allow the user to see a different "reality," a view which provides a different position relative to the scene without the user moving, but maintaining perspective and context, in order to expose or diminish other elements in the scene. Work in this area will help maintain privacy, especially in collaborative applications, or help to focus users' attention more narrowly on the main task in context.

Another active area of research of potential value is the occlusion of objects when rendering over an image. This technology could improve the feeling of immersion of the augmentation. The user's hand might appear over or on top of the augmentation, rather than being covered by the projected virtual image. Additional sensors for depth perception would be required in order to present a correctly occluded composition. Imminent mobile devices with stereo cameras may bring occlusion correction closer to reality.

The application of existing standards or their extensions for AR, on the sensor side in mobile devices requires more research on delivery of improved context for fixed objects. Additionally, processing of moving objects in the scene and variable light conditions needs to be improved. Improvements such as these, relying on both hardware and software, are currently the subject of research in many laboratories.

In a collaborative virtual or physical environment, there may be communicating hardware devices that interchange feature data (at a pixel level), annotations, or other abstractions of task specific data, in order to enhance creativity, capacity of shared spaces, resources, and stability. When networked, such collaborative devices could be provided to increase the perception of immersion and to provide a fluid and productive environment for collaboration.

These potential future AR applications will require deep interoperability and integration of sensors for data acquisition and presentation and significantly enhanced use of advanced silicon for imaging, video, graphics and compute acceleration while operating at battery-friendly power levels. Further, AR will benefit greatly from the emergence of interoperable standards within many divergent domains. Other standards topics that will need to be addressed by the community's collaboration with other domains in the future include rights management, security and privacy.

9   Conclusions

In this chapter we have discussed the current status of standards that can be used for interoperable and open AR, the issues and directions of development and use of international standards in AR applications and related services. We have identified the different standard bodies and players in different fields of interest involved in the development of AR. We have also analyzed the current state of standards within the mobile AR segment, specifically.

From widely available standards and the numerous potential applications, it is clear that for the industry to grow there must be further research to agree on standards, profiles suited to AR and for there to be discussion among AR experts on many different levels of development. While extending existing standards will be highly beneficial to achieve the ultimate objectives of the community, there must also be room for the inclusion of new ideas and evolving technologies. Therefore, standardization meetings for finding the best interconnection and synergies have emerged (i.e. International AR Standards Meetings).

These standards coordination meetings enable all players to not only discuss requirements, use cases, and issues but also the establishment of focused working groups that address specific AR standards issues and the generation and coordination of AR-related work items in the cooperating standards bodies. This approach fosters and enhances the process of standardization of AR in specialized fields in order that the community develops seamless and stable working products in the market.

References

1. ARWave Project Web page http://arwave.org
2. CityGML. 2008. OGC CityGML Encoding Standard. Open Geospatial Consortium, http://www.opengeospatial.org/standards/CityGML.
3. ECMA-262. 1999. Standard ECMA-262 3rd Edition. European Association for Standardizing Information and Communication Systems, http://www.ecma.ch/ecma1/STAND/ECMA-262. HTM KML. 2008. http://www.opengeospatial.org/standards/kml/
4. GML. 2006. OGC Geography Markup Language Encoding Standard 3.2.1. Open Geospatial Consortium, http://www.opengeospatial.org/standards/gml.
5. HTML5. 2011. Vocabulary and associated APIs for HTML and XHTML, http://dev.w3.org/html5/spec/.
6. KRONOS. 2008. *COLLADA 1.5.0 Specification.*
7. http://www.khronos.org/collada/
8. Pusta et al.. Spatial Relationship Patterns: Elements of Reusable Tracking and Calibration Systems, ISMAR '06 Proceedings of the 5th IEEE and ACM International Symposium on Mixed and Augmented Reality.
9. SOS. 2008. OGC Sensor Observation Service Interface Standard 1.0. Open Geospatial Consortium, http://www.opengeospatial.org/standards/sos.
10. WebGL. 2011. Open Khronos Group. WebGL Specification, http://www.khronos.org/webgl/specs/
11. WMS. 2007. OGC Web Map Service Interface Standard 1.3. Open Geospatial Consortium, http://www.opengeospatial.org/standards/wms.
12. WPS. 2009. OGC Web Processing Service Interface Standard. Open Geospatial Consortium, http://www.opengeospatial.org/standards/wps.

SmARt World - User-Friendly Mobile Ubiquitous Augmented Reality Framework

A.W.W. Yew, S.K. Ong, and A.Y.C. Nee

Abstract SmARt World is a framework for an ubiquitous computing smart space consisting of smart objects, which uses augmented reality (AR) to both create content in the smart space as well as interact with the smart space. The purpose of the framework is to support various mobile collaborative applications in an indoor AR environment. This chapter presents an early prototype of SmARt World, using a wireless mesh sensor network to facilitate smart objects, and an Android AR application for content creation and interaction with the smart space. A web-hosted server is used to maintain the smart space and allow multiple access from different users, thus facilitating a collaborative smart space.

1 Introduction

Augmented Reality (AR) allows computer generated graphics to overlay a real-life scene. By integrating virtual entities with real life, an AR interface allows simultaneous viewing and interactions of virtual and real objects. This capability is being applied to tasks like surgery [1], collaborative product design [2], entertaining and impactful presentations of augmented interactive graphics [3, 4, 5].

Due to the high computational requirements of AR applications, they have traditionally been targeted at desktop or wearable computers. Wearable computers are usually built around laptops and head-mounted displays. Sensors like inertial sensors and GPS receivers are often attached to wearable computers to track the user's location, e.g., the Tinmith System [6]. However, due to their weight and cumbersomeness, wearable computers have not reached widespread use. Now, recent

A.W.W. Yew (✉)
Department of Mechanical Engineering, Augmented Reality Lab,
National University of Singapore, EA 02-08,
National University of Singapore, 117576, Singapore
e-mail: andrew_yew@nus.edu.sg

L. Alem and W. Huang (eds.), *Recent Trends of Mobile Collaborative Augmented Reality Systems*, DOI 10.1007/978-1-4419-9845-3_3,
© Springer Science+Business Media, LLC 2011

advances in smartphone technology has seen vast improvements in computational power as well as embedded hardware and technological convergence, which has led to increased research and development in mobile AR, and great popular interest in mobile AR applications.

A novel idea has been to enable mobile AR applications to be collaborative, which greatly enhances the fun and usefulness of mobile AR. In Chapter 6 of this book, Vico et al. rightly identify mobile AR applications where live content is contributed by users as a very new and technically challenging area of research. One of the challenges of such applications is finding a way to allow end users more control over the content that they can contribute. Rather than be limited to just being able to enter text, audio recordings, or simple drawings, end users should be empowered to create richer AR content, in real-time and on an ad-hoc basis, without the need for technical knowledge in programming.

One limitation of many AR systems is, since virtual objects are simply overlaid on the real scene, users can only interact with virtual objects through the AR interface, but not with real objects. This limitation can be overcome by adopting a ubiquitous computing approach, where real objects are infused with intelligence and computing power so that they can be digitally connected to become smart objects and make the space containing these objects a smart space [7].

With the multitude of possible applications in AR, together with the advancing technology in mobile devices and ubiquitous computing, there is the motivation and tools to unify collaborative AR applications under a common framework that is accessible to the general public. This chapter presents such a unifying framework called SmARt World that aims to provide an AR platform which is easy to use and create content collaboratively and in real-time, and is extremely mobile for the user as only a smartphone is needed to interact with a smart space.

2 Related Work

There are currently a few AR application development frameworks, e.g., ARToolkit [8] and Studierstube [9], on which many applications have been built. These frameworks provide programming API for software developers to build AR applications based on fiduciary marker tracking. Some frameworks, e.g., the AR game creation system [10], hide the details of implementing the underlying AR API and only require the developers to focus on programming and creating content for the application. SmARt World aims to further abstract the application development by having a ready-made framework where applications can be created anytime and anywhere by anyone without having to do any programming or scripting.

There are AR systems that are related to ubiquitous computing. The Ubiquitous Augmented Reality (UAR) system [11] utilizes a hybrid vision-inertial tracking method for achieving registration of virtual objects to the real environment, and a wireless sensor mesh network that facilitates multiple modes of environment sensing, with the augmented world viewed using a wearable computer. While the architecture of the SmARt World framework is similar to UAR, SmARt World presents an improvement on the user side by developing a user-friendly AR interface, making it easy for content creation and interaction with the smart space using just a widely available, light and portable smartphone.

3 System Design and Implementation

Some common and basic requirements of most AR systems are achieved in this framework, namely registration of the virtual models to the environment, graphics rendering, and user interaction with the augmented world. For collaborative applications, multiple users must be connected to a single instance of the augmented world. Furthermore, this system is targeted at the general public for content creation as well as end-users, so it must be user-friendly.

The SmARt World framework has a three-layered architecture – physical layer, middle layer, and AR layer (Figure 1). The physical layer consists of a wireless mesh network of smart objects. The middle layer consists of a single server to process and manage data from the physical layer and the smart space, and to relay commands between the physical layer and the AR layer. The AR layer is the smartphone running the AR application to view and interact with the smart space.

3.1 Physical Layer

The physical layer is a wireless mesh network of smart objects. The smart objects are nodes on the mesh network which are all connected to a root node either directly or via multiple hops through each other. The root node has a hard USB link to the server in the middle layer.

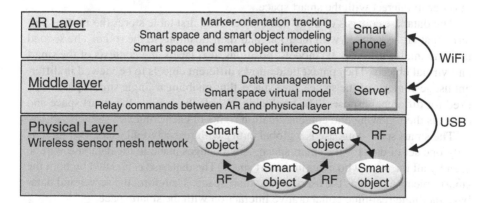

Fig. 1 SmARt World system architecture

Smart objects are simply ordinary objects each attached with an RF transceiver with a unique ID which communicate on the mesh network. The RF transceiver can either be connected to sensors or wired to electrical components. Since communication between the mesh network and the server is two-way, the smart space is able to provide the functions of both environment sensing as well as remote control of the smart space and the objects contained therein.

In the prototype, an RF transceiver is made the root node of the mesh network by connecting it to a PC by USB. The USB connection both powers the RF Engine and links up with the server by a virtual serial port connection. The root node is uploaded with a script to ping the mesh network from time to time to detect new nodes or lost nodes, listen for data coming from the nodes in the mesh network, and send data to the server to update the database with data from the nodes. The RF transceivers of the other nodes are battery-powered. The general purpose pins on the RF transceivers can be attached with sensors or wired to electronics to control them. Scripts must be written and uploaded to the RF transceivers to send and receive signals to and from the devices that the RF transceivers are attached to. Much convenience is brought by the RF transceivers' ability to form and heal the mesh network automatically as smart objects are brought in and out of range of each other. This ability of the network to heal automatically ensures that the digital existence of real objects does not disrupt the normal usage of them when they are moved about.

3.2 Middle Layer

The middle layer is a software layer which performs database management for the smart space, smart objects and users in the smart space, and links the physical layer with the AR layer. It consists of a web-server to interface with the smartphone, an SQL server for database management, and Java program to communicate with the mesh network. The smartphone only needs an Internet connection to communicate with the web-server, which stores and retrieves data to and from the SQL database, in order to interact with the smart space.

The database consists of three main tables. The first table stores the marker patterns that are associated with the room and their positions in the room. The second table maintains the data, 3D models, positions, text labels and contexts of the smart and virtual objects. The context field allows different objects to be viewed in different usage contexts of the same smart space, thus enabling a single smart space to be used for different purposes. The final table records the users in the smart space and controls their viewing and content creation rights of each context.

The smart space database is a global entity with respect to all users, i.e., there is only one set of data for the smart space. Any changes made to the database will be seen by all the users who enter the smart space. The database is updated by both the smart objects as well as the users in the smart space. Therefore, the server and database facilitate real-time collaborative interaction with the smart space.

3.3 AR Layer

The AR layer consists of the smartphone with the AR viewing and interaction application running on it. There are three main tasks for the AR layer – registration of virtual objects to the real space of the user's surroundings, rendering all virtual objects, and interaction with virtual and real smart objects.

3.3.1 Tracking and Registration

The tracking and registration method employed uses a combination of marker-tracking and orientation sensing. A fiduciary marker is used to establish registration of the virtual model of a smart space to the actual smart space. The orientation reading of the embedded orientation sensors when a detected marker is taken as an initial orientation, with subsequent orientation changes tracked when the marker goes out of view. The orientation changes are applied to the 3D graphics renderer to transform the virtual objects (Figure 2).

3.3.2 Rendering Virtual Objects

Virtual objects are rendered in the 3D virtual coordinate frame of the smart space, which is registered to the physical smart space using the marker-orientation tracking method described previously.

A feature of the virtual objects in the SmARt World is that these virtual objects are affected by real objects in various ways, of which two have been implemented – occlusion (Figure 3) and collision. Occlusion of virtual objects by real objects is important in applications like navigation, where the user needs some indications of the relative depth between the augmented graphics and the real scene. Collision between objects will be useful when the smart space is used for games involving a virtual object to be interacted with by real objects.

For real objects to have an effect on virtual objects, virtual models of these real objects are needed to calculate the regions on the virtual objects that are occluded. In the present implementation, real objects are modeled using bounding boxes. When a bounding box is in front of a virtual object in the 3D smart space, only the non-occluded part of the virtual objects will be displayed.

A simple method of testing the vertices of each virtual object against the bounding boxes of real objects and of each other is used to detect collisions. When a collision is detected, the virtual object's movement will be restricted.

3.3.3 Interaction with the Smart Space

All interactions with the smart space are achieved through the touch-screen interface of the smartphone.

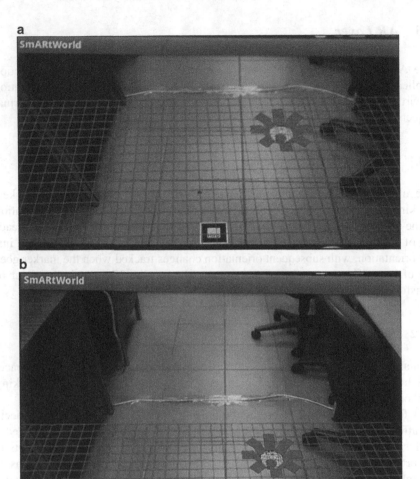

Fig. 2 Marker-orientation tracking where (**a**) shows registration using a fiduciary marker and (**b**) shows the registration maintained using orientation sensors

Virtual objects can be selected by tapping on them on the touch-screen of the smartphone. Selected virtual objects can then be moved around the smart space by selecting and dragging them across the touch-screen.

The smartphone menu is used to choose the modes of control of smart or virtual objects and change the rendering options, e.g., toggle rendering of the ground and bounding boxes. The menu is context sensitive, i.e., depending on the type of object selected, different menu options are generated. If a smart object is selected, menu options to control the smart object and add text annotations will appear. If no object is selected, the menu allows the users to use the touch-screen to add bounding boxes, and position and scale them to coincide with real objects.

Fig. 3 A real table occludes a virtual flower

4 System Application

The system is designed to make it easy for content creators as well as end-users, who play a large part in collaborative content creation. This section describes the process of setting up and using a smart space under the SmARt World framework.

4.1 Smart Objects

The attachment of RF transceivers to real objects makes these objects smart objects. An RF transceiver is battery-powered and wired to electronics. Programming the RF transceiver requires scripting knowledge, and interfacing them with devices requires different modes of usage of the I/O pins of the transceiver. For example, if a device is to be controlled by an RF transceiver, general purpose pins on the transceiver are connected to the device and the pins set as output. To obtain data from sensors, the general purpose pins will be set as input.

Once the RF transceivers are properly attached to the real objects, no further work is needed. Smart objects will automatically be handled by the server and controlled through the AR interface on the smartphone. Users of the final SmARt World framework are not expected to build smart objects as these objects would be assembled as smart objects by the manufacturers in the future.

4.2 Modeling the Smart Space

Modeling the smart space involves positioning and sizing 3D bounding boxes around real objects. This is done using the touch-screen interface on the smartphone.

The first step is to choose an origin for the virtual model of the smart space in the physical space and place the fiduciary marker. The pattern of the marker is stored in the server and its x-y-z-coordinates are (0,0,0) in the virtual smart space. The next step is to use the AR interface on the smartphone to add, position, and scale the bounding boxes around real objects in the smart space, and add text annotations to each bounding box.

4.3   Interaction

Interaction with the smart space is achieved through the same interface as the modeling of the smart space. The same controls are available to add and manipulate objects. The virtual smart space starts rendering once a marker is detected. Multiple users, through their own smartphones, can look around the smart space, touch objects to select and move them, and display and edit text annotations. This basic functionality forms the basis for any AR application that uses SmARt World as its framework.

5   Prototype Testing

5.1   Procedure

A prototype of the SmARt World was implemented using the authors' laboratory as the smart space in order to test the framework. Synapse RF Engines were used as RF transceivers. The server and SQL database were hosted on a desktop computer, and the AR interface was implemented as an Android 2.1 application running on a Samsung Galaxy S smartphone. As an end-user who is only visiting the smart space, only a smartphone with a wireless Internet connection is needed.

First, an RF transceiver was used as the root node of the mesh network by connecting it to a PC by USB to link to the server. The root node was controlled by a script to ping the mesh network from time to time to detect nodes in range, listen for data coming from nodes in the mesh network, and send data to the server to update the database. Two RF transceivers were attached to smart objects. One RF transceiver had an LED attached and a script to toggle the LED on or off depending on the incoming signal from the server. The other RF transceiver had a humidity sensor attached and a script to update the server on the sensed humidity level.

The system was first tested from the viewpoint of a smart space creator. The AR interface on the smartphone was used to model the laboratory by drawing bounding boxes around a table, cabinet, fan, as well as the two smart objects that are built with RF Engines. A virtual model of a flower was inserted into the smart space as well. Next the system was tested from the viewpoint of a smart space visitor. The visitor

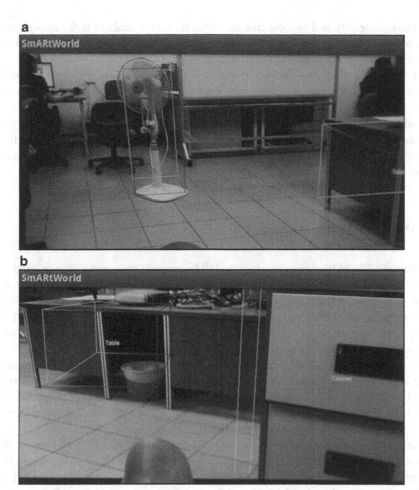

Fig. 4 Screenshots of the laboratory modeled as a smart space

registered the virtual smart space by looking at a fiduciary marker, and then looked around the room. He selected the virtual flower and moved it around the room, demonstrating occlusion by and collision with the table, cabinet and fan. The LED on a smart object was turned on and off from the smartphone, and humidity level from the other smart object was read.

5.2 Evaluation of the Prototype

The tests of the prototype show that modeling of the smart space using the touch-screen of the smartphone is fast and easy. Virtual 3D models which can collide with and be occluded by real objects give an added feeling of realism to the experience, which makes it more immersive. Furthermore, using the AR and touch screen interface to control smart objects is intuitive.

However, the marker-orientation tracking is slightly sluggish and slight accuracy is lost when the orientation sensors take over from the marker tracking to maintain registration. It was found that, while the occlusion and collision between real and virtual objects added a feeling of realism to the experience, the reduced accuracy of the marker-orientation tracking coupled with the limited details in which bounding boxes could represent real objects resulted in some perceivable defects in the overall effect of occlusion and collision.

6   Possible Applications

The SmARt World framework is designed to be suitable for many different applications. Without closing the smartphone application, a user could move from one smart space to another, changing contexts between smart spaces and within the same smart space. Using the same database, the context of a smart space could be changed by only augmenting the context-related objects to the smart space. In this section, some possible applications for different contexts are discussed.

To use the SmARt World framework for museum touring and guidance, each exhibit area could be turned into a smart space with relevant virtual models inserted to add interactivity and collaborative story-telling to the exhibition piece, annotations added by curators to provide more information, and annotations left by museum visitors as comments or contributions of information.

Another possible application is an assistive smart space, particularly for walking-impaired patients. Their room can be turned into a smart space where the various electrical appliances, e.g., television, lights, and air-conditioner, could be made into smart objects which the patient can control using the AR interface. Furthermore, the patient can beautify it with virtual objects and have visitors join in with the decorating and use virtual objects to play games with the patients.

Public places like shopping malls are good candidates for multiple contexts per smart space (Figure 5). One context is for navigation to guide shoppers to specific shops or facilities. Another context can add some entertainment in the form of interactive virtual objects which shoppers can interact with collaboratively.

7   Future Work

SmARt World is still early in development. The approaches in implementing some of the basic features like marker-orientation tracking and modeling real objects using bounding boxes are not ideal. Some basic features like position tracking of smart objects are also not completed.

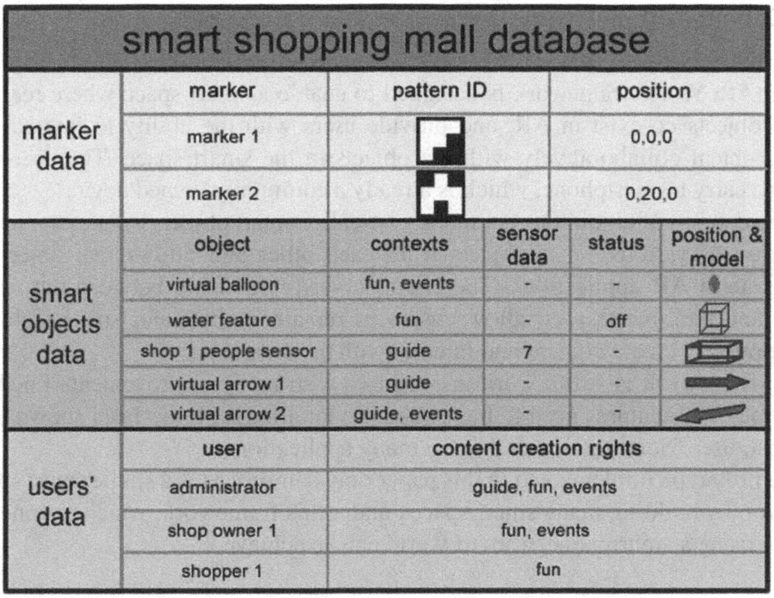

smart shopping mall database				
marker data	marker	pattern ID		position
	marker 1	◼		0,0,0
	marker 2	◼		0,20,0

smart objects data	object	contexts	sensor data	status	position & model
	virtual balloon	fun, events	-	-	◆
	water feature	fun	-	off	▢
	shop 1 people sensor	guide	7	-	▭
	virtual arrow 1	guide	-	-	➡
	virtual arrow 2	guide, events	-	-	⬅

users' data	user	content creation rights
	administrator	guide, fun, events
	shop owner 1	fun, events
	shopper 1	fun

Fig. 5 Example database for a shopping mall with multiple contexts

An alternative to the marker-orientation tracking approach is markerless tracking. There has been recent research on this such as the work by Fong et al. [12] that uses natural planar surfaces as markers, and PTAM [13] which uses computer vision to build a 3D map of a small workspace. These two systems enable AR without fiducials or a prior map.

The bounding box approach to modeling real objects is not ideal for realistic interaction with virtual objects. In a larger environment such as a supermarket, it could also become more difficult than what was experienced in the prototype test. There has been work by Lee et al. [14] on 3D modeling of real objects using a mobile device and a server to automatically create 3D models from photographs with some user annotation.

One significant limitation of the current prototype system is the lack of real-time position tracking of smart objects, thereby only allowing stationary smart objects. There are some experimental indoor localization methods based on Received Signal Strength Indicator (RSSI), with one such method utilizing a combination of RSSI and RFID reference tags to reach an accuracy of 0.45 metres [15].

Finally, a larger-scale prototype test is planned involving multiple users in a more practical scenario, such as a supermarket, with an administrative context for monitoring conditions like freezer temperatures, and a customer context for aiding in the locating of various sections in the supermarket. Before this larger-scale test, further refinement will be done to deal with the afore-mentioned problems.

8 Conclusion

The SmARt World framework is designed to enable a smart space where real and virtual objects co-exist in AR, and provide users with the ability to interact and create content collaboratively with the objects in the smart space. The user only needs to carry a smartphone, which is already a commonly owned device

The adoption of ubiquitous computing to realize smart objects isolates the implementation of various smart objects from each other and allows real objects to participate in AR applications. The client-server relationship between the smartphone and the smart space allows users to remain mobile and still be able to collaboratively create content and interact with the smart space.

The framework is still a work-in-progress, with many improvements lined up, and some basic features need to be revised. So far, though, it has been shown to be low-cost, user-friendly and suitable for many applications.

The prototype implemented in this paper only demonstrates a single smart space. However, by building many smart spaces under this framework, which anyone can explore using a smartphone, a smart world can be achieved.

References

1. Paloc C, Carrasco E, Macia I, Gomez R, Barandiaran I, Jimenez JM, Rueda O, Ortiz de Urbina J, Valdivieso A, Sakas G (2004) Computer-aided surgery based on auto-stereoscopic augmented reality. Proceedings of the Eighth International Conference on Information Visualisation pp. 189–193. 14–16 July 2004. doi: 10.1109/IV.2004.1320143
2. Shen Y, Ong SK, Nee AYC (2008) Product information visualization and augmentation in collaborative design. Computer-Aided Design 40(9):963-974. doi: 10.1016/j.cad.2008.07.003
3. Grafe M, Wortmann R, Westphal H (2002) AR-based interactive exploration of a museum exhibit. The First IEEE International Workshop on Augmented Reality Toolkit pp. 5. 6 January 2003. doi: 10.1109/ART.2002.1106945
4. Miyashita T, Meier P, Tachikawa T, Orlic S, Eble T, Scholz V, Gapel A, Gerl O, Arnaudov S, Lieberknecht S (2008) An Augmented Reality museum guide. ISMAR 2008 pp. 103–106. 15–18 September 2008. doi: 10.1109/ISMAR.2008.4637334
5. Hurwitz A, Jeffs A (2009) EYEPLY: Baseball proof of concept – Mobile augmentation for entertainment and shopping venues. ISMAR-AMH 2009 pp. 55–56. 19–22 October 2009. doi: 10.1109/ISMAR-AMH.2009.5336723
6. Piekarski W (2006) 3D modeling with the Tinmith mobile outdoor augmented reality system. Computer Graphics and Applications, IEEE 26(1):14-17. doi: 10.1109/MCG.2006.3
7. Ma J, Yang LT, Apduhan BO, Huang R, Barolli L, Takizawa M, Shih TK (2005) A walk-through from smart spaces to smart hyperspaces towards a smart world with ubiquitous intelligence. Proceedings of the 11th International Conference on Parallel and Distributed Systems pp. 370 – 376. 20 – 22 July 2005. doi: 10.1109/ICPADS.2005.60
8. HIT Lab. ARToolKit. Retrieved 17 August 2010 from http://www.hitl.washington.edu/artoolkit
9. Studierstube. Studierstube augmented reality project. Retrieved 25 August 2010 from Studierstube.icg.tu-graz.ac.at/index.php
10. Wilczyhski L, Marasek K (2007) System for Creating Games in Augmented Environments. International Conference on Multimedia and Ubiquitous Engineering pp. 926-931. 26–28 April 2007. doi: 10.1109/MUE.2007.200

11. Li X, Chen D, Xiahou S (2009) Ubiquitous Augmented Reality System. Second International Symposium on Knowledge Acquisition and Modeling 3:91–94. doi: 10.1109/KAM.2009.312
12. Fong WT, Ong SK, Nee AYC (2009) Computer Vision Centric Hybrid Tracking for Augmented Reality in Outdoor Urban Environments. 16th Symposium on Virtual Reality Software and Technology pp. 185-190. 14–15 December 2009. doi: 10.1145/1670252.1670292
13. Klein G, Murray D (2009) Parallel Tracking and Mapping on a camera phone. ISMAR 2009 pp. 83–86. 19–22 October 2009. doi: 10.1109/ISMAR.2009.5336495
14. Lee W, Kim K, Woo W (2009) Mobile phone-based 3D modeling framework for instant interaction. IEEE 12th International Conference on Computer Vision Workshops pp. 1755-1762. 27 September–4 October 2009. doi: 10.1109/ICCVW.2009.5457495
15. Zhang D, Yang Y, Cheng D, Liu S, Ni LM (2010) COCKTAIL: An RF-Based Hybrid Approach for Indoor Localization. IEEE International Conference on Communications pp. 1-5. 23–27 May 2010. doi: 10.1109/ICC.2010.5502137

Augmented Viewport: Towards precise manipulation at a distance for outdoor augmented reality wearable computers

Thuong N. Hoang and Bruce H. Thomas

Abstract In this chapter we present our research directions on the problem of action at a distance in outdoor augmented reality using wearable computers. Our most recent work presents the augmented viewport technique to enable interactions with distant virtual objects in augmented reality. Our technique utilizes physical cameras to provide real world information from the distant location. We examine a number of factors required to achieve an efficient and effective solution for precise manipulation at a distance for outdoor augmented reality using wearable computers. These include the improved usage of physical cameras, collaboration, viewpoint visualization, alignment of virtual objects, and improved input devices. We particularly focus on the collaboration aspect of the technique, with the utilization of remote cameras from multiple users of wearable computer systems and mobile devices. Such collaboration supports precise manipulation tasks by allowing viewing from different perspectives, directions, and angles, as well as collaborative precise triangulation of virtual objects in the augmented environment.

1 Introduction

Our investigations into augmented reality (AR) have focused on wearable computers in an outdoor setting [1]. We are motivated to find solutions to the problem of precise action at a distance for outdoor AR [2]. One of the main challenges is the requirement of mobility. Users of an outdoor wearable computer system require freedom of movement and wearability comfort. Indoor tracking systems with

T.N. Hoang (✉)
Wearable Computer Lab, School of Computer and Information Science,
University of South Australia, Mawson Lakes Campus, 1 Mawson Lakes Boulevard,
Mawson Lakes, SA 5010, Australia
e-mail: contact@thuonghoang.com

L. Alem and W. Huang (eds.), *Recent Trends of Mobile Collaborative Augmented Reality Systems*, DOI 10.1007/978-1-4419-9845-3_4,
© Springer Science+Business Media, LLC 2011

complex setups can provide high precision operations, such as the FastTrak's Polhemus magnetic tracking [3] or the Vicon motion system for visual tracking [4]. However, such solutions cannot be applied outdoors because they restrict user's movements and require complex setup at a fixed location. The main environmental constraints of the outdoor environment are its inherently large scale and dynamic nature. Therefore, it is common for users to interact with virtual objects that are out of arm's reach. Action at a distance (AAAD) technique is one of the approaches of handling these outdoor constraints. Solutions to the AAAD overcome these environmental constraints and offer the users of outdoor AR systems an efficient and effective way of interacting with distant virtual objects in the augmented outdoor environment in a collaborative manner.

1.1   Action at a distance problem

AAAD is a well researched problem in the VR domain. The two main approaches are: bringing distant objects closer to the user and bringing the user closer to distant objects. There are many interaction techniques belonging to those categories, such as: world-in-miniature, voodoo doll, image plane techniques, and ray casting, cone selection, Go-Go arm, HOMER, and teleportation techniques.

World-in-miniature (WIM) [5] is an interaction technique that supports multiple interactions including object selection, manipulation, navigation, and visualization. The virtual world and its contents are scaled down into a miniature version and placed on one of the user's virtual hands. By manipulating the miniature objects in the WIM, the user is able to change the position, orientation, and scale of the corresponding objects in the virtual environment. The user may also navigate around the virtual environment by manipulating their representations in the virtual world [6]. With a similar scaling approach, the scaled world grab technique brings the world closer to the user, by scaling the world centered to the user's head [7].

A variation of this approach is to scale down only the distant object to be manipulated. The Voodoo doll technique [8] places a miniature version of the distant object of interest in the user's hand for manipulation. The two-handed technique supports translation and rotation of virtual objects by performing relative movements and rotations with two Voodoo dolls on both hands. The creation of Voodoo doll uses the image plane technique by pinching the projection of distant objects on the user's viewing plane. The image plane technique enables interaction with the projection of the world. Distant objects are projected onto the user's two-dimensional viewing plane [9]. Image plane interaction supports various gestural selection techniques, such as pinching, pointing, and framing, as well as affine transformations, such as translation, scale, and rotation [10].

A second approach of covering the distance between the user and the virtual objects is to bring the user closer to the objects. Ray casting and cone techniques extend the user's pointing gesture for selection of distant object. A virtual ray is fired from the user's virtual hand or pointer to select distant objects [11]. These techniques are suitable for object selection from both a close range and at a distance. The user can translate the object by moving the virtual ray to change the position of the object which is attached at the end of the ray. Extending arms is another popular approach for bringing parts of the user's body to distant objects. The virtual hand metaphor is an intuitive technique supporting direct manipulation of virtual objects with the user's hand. A simple version of the technique implements a one-to-one mapping of the movements between the user's hands and the virtual ones, limiting it to close body interactions [12]. The Go-Go arm technique [13] allows the user to extend their virtual hands by non-linearly increasing the mapping ratio as the physical hand reaches away from the body. The HOMER technique [14] is a combination of both ray casting and hand extension techniques that allows a seamless transition between selection and manipulation tasks.

Teleportation instantly transports the user to remote locations in the virtual environment in order to perform close body interaction with distant objects. This technique is typically considered a travel technique in virtual reality [15], which could be utilized to overcome distance in the task of manipulating distant virtual objects. Teleportation has the visual advantage over hand extension techniques, as the user visually interacts with distant objects at a close range.

1.2 *Augmented viewport technique*

Our recent work presents the augmented viewport technique [2], a new AAAD technique for outdoor AR systems. The augmented viewport is the AR version of the virtual viewport technique in VR research [16], which brings a view of a distant location closer to the user. The augmented viewport takes the form of a virtual window, showing both the virtual and real world information of a distant location. The real world information is provided by physical cameras, which are either 1) located at the distance outside or 2) within the user's field of view, or 3) located near the user and equipped with an optical zoom lens to zoom further into the distance. Fig.1 shows an example of the augmented viewport using a zoom lens camera. The camera is mounted on a tripod and zoomed in on a physical window on a building at a distance (Fig. 1a). The inset figure is the view as seen through the zoom camera. The view for the user without the augmented viewport through the standard head mounted display (Fig. 1b) indicates virtual objects (blue boxes) overlaying on the same window of the physical building. Combining those views results in an augmented viewport with a window display, showing distant virtual and physical objects. The viewport offers a closer view, and is located near the user (Fig. 1c). Currently the augmented viewport technique is implemented on the Tinmith wearable computer system. Tinmith is an outdoor AR wearable computer system [1], consisting of a belt-mounted computer and a wearable helmet. Tinmith employs a video see-through head mounted display, located on the helmet together with a head orientation sensor and a GPS antenna unit.

The outdoor environment is naturally large and open. The augmented viewport technique overcomes this aspect of the outdoor environment and enables the user to perform direct manipulation operations on distant virtual objects. The user is able to interact with distant virtual objects through the viewport using close body interaction techniques, such as ray casting or image plane techniques. The possible placements of the augmented viewport are in three coordinate systems relative to the

Fig. 1 The augmented viewport technique (**a**) The zoom lens camera setup on tripod. Inset: the view through the zoom camera of a distant physical window (**b**) The view through the user's head-mounted display (blue boxes indicating distant virtual objects overlaying the physical window) (**c**) = (**a**) + (**b**): The augmented viewport, combining the video stream from the zoom lens camera and distant virtual objects

Fig. 1 (continued)

user's view: world, body, and head relative. World relative placement has the view-port in a fixed position and orientation in the world coordinate system. Body relative places the augmented viewport attached to the user's body, so that the viewport stays at a fixed position and orientation from the location of the user's body. And head relative placement fixes the viewport to the user's head orientation.

1.3   Collaboration for action at a distance

The augmented viewport technique supports collaboration for action at a distance problem. User of the augmented viewport technique can benefit from the head-mounted or mobile cameras from other users of wearable computer systems and mobile devices. A network of mobile devices and wearable computers are intercon-nected to provide multiple viewpoints of the augmented environment. Different viewpoints can be shown through the augmented viewport windows to assist with manipulation tasks of virtual objects at a distance.

There can be several camera sources to form a network. Head-mounted cameras worn by users' of wearable computer show direct first person perspectives of the world. Cameras attached to mobile devices can be used to offer more flexible and dynamic viewing angles. Outdoor wearable computer user can also extend manipu-lation to indoor settings with desktop webcams.

The collaboration network of remote cameras offers various benefits. The user has multiple viewing angles, directions, and aspects of the environment, including the views that physically impossible for other users to achieve. Triangulation of virtual objects can be achieved from multiple camera sources, in order to precisely determine virtual objects' positions.

The augmented viewport is our first step towards achieving the research goal of precise manipulation techniques for AAAD in outdoor AR environments. For future research directions, we analyze various factors that are posing challenges to the open research question of AAAD for outdoor AR.

2   Required factors

The required factors to improve AAAD technique for outdoor augmented reality are fivefold: 1) improved usage of remote cameras, 2) collaboration between users of wearable computers and mobile devices, 3) better visualization of views from these cameras, 4) the automatic alignment (snapping) of virtual objects to physical objects, and 5) improved user input devices.

2.1   Remote cameras

The augmented viewport technique can be applied to different types of cameras, such as CCTV cameras, remote wireless cameras, or even static imagery from satellites. We are also interested in the collaborative potential of using the AR cameras from multiple users of nearby AR systems, and/or cameras mounted on automatic robots. There is a requirement to better perform the discovery and selection of existing cameras in the user's world, as well as to understand their locations and placements.

A CCTV camera provides a fixed viewpoint at a particular distant location. If the camera is located at a location that is not visible to the user, possibly due to physical constraints, the augmented viewport offers a telepresence mode of interaction. Considering that the viewport appears as a virtual window not covering the entire field of view, the user still retains their physical presence in the augmented world, which is one of the main characteristics of the outdoor AR system. We propose increasing the flexibility to the augmented viewport solution by providing the user with the ability to control the orientation and/or the position of the remote camera. ARDrone[1] is a wireless controlled flying quadrotor, with a VGA camera mounted at the front. The ARDrone can be connected to via an ad-hoc wireless link, through which the flying commands and the video stream of the front-mounted camera are transmitted. A potential implementation can have the user control the ARDrone with either an onscreen menu or a separate orientation sensor attached to the hand.

---

[1] http://ardrone.parrot.com

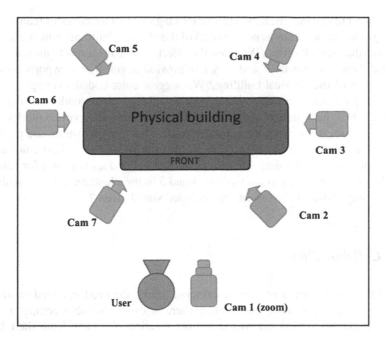

Fig. 2 Camera viewpoint scenario

This allows the user to remotely control the ARDrone and to use the video source from the camera as an augmented viewport. With this combination, the user can selectively perform manipulation of virtual objects from various distant locations. A similar source of remote cameras are remote controlled robotic vehicles, which have been utilized in an AR X-Ray system [17].

Static satellite imagery has been shown to improve precision of virtual object placement. The work by Wither and colleagues [18] overlays static satellite images of the user's surroundings in a God's eye immersive view, in order to assist the user with the placements of virtual annotations. The augmented viewport can take advantage of this approach and be used to provide a top-down map view of the user's location, while still allowing the user to maintain the immersive view of the augmented world.

Considering that there are various sources of cameras for the augmented viewport technique, a question arises as to how an AR system performs the discovery of available cameras in the vicinity? How to properly present their existence, as well as their locations and orientations, to the user in order to make a selection of cameras for the manipulation task? Will a top-down map view of the vicinity suffice to provide effective visualization of the camera cluster? Each camera is characterized by their location in the physical world and their intrinsic parameters, especially their viewing frustum. Fig. 2 outlines a scenario for possible locations of physical cameras. Camera no. 1 has a zoom lens to zoom closer to distant location while situated near the user, and cameras no. 2 and 7 are looking from similar views to the user's perspective. Other cameras, however, are looking from either the sides (cameras no. 3 and 6) or the back (cameras no. 4 and 5) of the physical building, which cannot be seen from the user's location. What are the effective visualizations to indicate to the user that those cameras exist and they can provide augmented viewports covering certain areas of the physical building? We suggest color-coded shading, in which colored regions are overlaid on the physical building corresponding to the areas seen by each of the remote cameras. For each remote camera, a virtual model of the viewing frustum is created with separate colors. The virtual frustums are placed at the locations of the respective cameras, so that the frustums can cast color-coded shades onto physical building. However, this technique does not cater for cameras in occluded regions, such as camera no. 4 and 5 in the illustration. Is it possible to utilize X-Ray vision [17] to assist the occluded visualization?

2.2 Collaboration

One of the many sources of remote cameras can be the head-mounted camera of another wearable computer user. When there is other wearable computer users (*remote users*) located closer to the remote location, the view from their head-mounted cameras be displayed in augmented viewports of the wearable computer user who is located further away from the desired location (*local user*). This setup supports collaborative virtual object manipulation, in which the *remote user* can directly manipulate virtual objects through the wearable computer interface, and the *local user* can perform manipulation through the augmented viewport.

An example scenario is when multiple architects are examining a large architectural site with virtual models of buildings and/or structures. The architects can collaboratively manipulate parts of the models while observing from different viewpoints and locations. Such a collaboration feature is also useful in situations where the participants cannot visually communicate with one another, such as a collaborative repair or inspection scenario in outer space.

An important factor to be considered for collaboration is the support of simultaneous manipulation. There needs to be a sharing mechanism to prevent multiple users from interacting with the same virtual object at the same time. When a user is manipulating the virtual objects, extra visualizations, such as virtual hands or onscreen cursors depicting their current operations, can be shown to other users in the network.

2.3 Viewpoints

Once the availability and the locations of remote cameras have been presented to the user, the next step is to select a suitable camera to perform the required manipulation task. What are the selection criteria to? The following factors should be taken into consideration regarding their effects on task performance: the angle the camera is looking at the distant area of interest, the level of details provided by the camera, and whether the user can intuitively comprehend the alignment between their viewpoint and the camera's viewpoint. The extent to which these factors affect task performance is still open to further research. A study by Zimmons and Panter [19] suggests that the quality of the rendered virtual environment does not affect the user's task performance, possibly due to the physiological condition of threat to personal safety, which was set up in the study.

There is a requirement to better understand how the augmented viewpoints from remote cameras align with the user's physical view of the world. The effects of the misalignment between the viewpoints of the remote cameras and the users on their task performance and sense of presence remains an open research question. In outdoor AR systems, it is important for the user to maintain the first person perspective throughout. Remote cameras, however, may be looking at different angles and directions. For example, the user is currently looking directly in front of a physical building and the augmented viewport uses a CCTV camera pointing to the side or the back of the same building, completely out of the user's field of view. This scenario may affect the task performance. If the remote cameras and the user are looking in opposite directions, such as cameras no. 4 and 5 in Figure 2, the translation task will be affected by the mirror effect, such that when the user translates the object to the left through the viewport, the object ends up moving to the right. How can such errors be prevented? What are effective techniques to help the users visualize the locations and directions of the remote cameras, especially when they are different from the user's perspective? One suggestion is to perform ray tracing technique to select camera. A set of rays are cast from the location of interest towards the camera cluster to determine which camera can provide coverage for that area. However, this technique requires a virtual model of the physical building to determine the area that the cameras are viewing.

Regarding the viewpoint misalignment, we plan to apply planar homographies to generate a view of the distant location from multiple remote cameras so that it matches with the user's first person perspective. We expect that this approach will enhance the intuitiveness of the augmented viewport, but this must be evaluated.

Multiple viewport interaction is an interesting research topic regarding the viewport window interface in VR. The work by Hirose and colleagues [20] supports object manipulation and user navigation through the use of multiple viewport windows. Virtual objects can be translated from one remote location to another by travelling through the viewports that are showing the corresponding remote locations. For the augmented viewport, a question arises as to how the user can effectively interact with multiple viewports from multiple physical remote cameras. The cameras show either different angles of a single remote location or separate remote locations.

2.4   Precision by snapping

We also question the precision aspect of the AAAD problem. What does precision mean in the context of an AR environment? One of the most important aims in AR

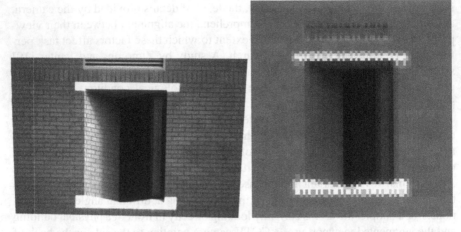

Fig. 3 Pixel granularity: Top: seen through the viewport. Bottom: seen through the normal HMD on the wearable computer system. Both are pointing at the same physical window

is to align the physical world and the virtual information, and merge them into a unified reality. Therefore, precision is said to be achieved when virtual objects are correctly aligned with the physical worlds and/or other virtual objects. Currently, sensor and tracker errors are still existent. What are the approaches to further improve precision, within the limitations of the current state-of-the-art registration and tracking technologies? We seek to reduce freehand errors and reinforce physical-virtual object alignments.

Snapping is a proven technique for improving precision by reducing freehand errors, as widely implemented in desktop-based CAD applications [21]. A typical implementation of snapping is grid-based snapping. It is an open question as to the proper approach of rendering a grid-based visualization in an outdoor augmented reality environment to improve precision in virtual object manipulation. Such rendering may be obtrusive and could interfere with the tasks.

A pixel is the smallest displayable element on the head mounted display. The larger the distance from the camera to the physical object, the fewer the number of pixels the physical object occupies on the screen. The augmented viewport uses cameras that can have a closer view of the object, and increases the number of pixels representing the object on the head mounted display; thus increasing the granularity in precision for the task of aligning virtual objects to the physical world. Fig.3 illustrates the difference in pixel granularity between the zoom lens camera and the HMD camera. Both images show the same physical window, with the left hand side being seen through the viewport and the right hand side through the HMD camera. With the viewport, the top and bottom white edges of the window can be clearly distinguished, while they are very blurry in the HMD camera. We propose using feature and edge snapping to improve the alignment between physical and virtual objects, through the pixels of the video stream of the augmented viewport. A selected image-based edge and feature detection algorithm is applied to the video stream provided from the remote cameras and appropriate visual cues are displayed through the augmented viewport. User's manipulations of the virtual objects can be snapped to detected features or edges, if required. Once the virtual objects are aligned to the features in the video stream, it is important to understand the correlation in transformation between the remote cameras and the physical world. What are the position and direction of the camera relative to the world? How does such information affect the proper alignments between the virtual objects and the physical world?

Snapping improves the alignment between virtual objects and the video stream from the augmented viewport (Fig.4a). Through camera discovery, the intrinsic parameters of the physical camera are known, as well as its locations relative to the physical world (Fig.4b). Combining both improvements will lead to a more precise alignment between the virtual object and the physical world at a distance. Fig.4c illustrates such improvements with more precise alignment of the virtual blue boxes to a distant physical window. The process involves different coordinate systems. The first step is the alignment of the virtual objects within the window coordinate system of the augmented viewport. The second phase is to translate the position of the virtual objects into the global world coordinate system, based on the alignment of the contents of the augmented viewport within the global coordinate system. Lastly, the process is concerned with how to precisely render the location of the virtual object to view in the user's head-mounted display. This last stage includes the position and head orientation of the user within the global coordinate system.

Such improvements also lead to another approach of reducing free hand errors by specifying the physical-virtual objects alignment as constraints for manipulation tasks. Once a virtual object has been aligned to a particular feature in the physical world, it can be set to be constrained to that feature throughout subsequent manipulation operations. For example, a virtual window is constrained to the left vertical edge of a physical window, so that any subsequent translations can only move the virtual window along the vertical edge.

Further precision can be achieved through the application of image processing. Considering that the location of the user is known through the GPS tracker, and the position, orientation, and the intrinsic parameters of the remote cameras are obtainable, it is possible to perform image-based distance measurements. A mapping is formed between the pixels on the augmented viewport and the actual distance measurements of the physical features shown in the viewport. Therefore, based on the number of pixels the virtual objects occupy on the screen, we can specify or obtain the measurements of the virtual objects, or use known measurements to set constraints for manipulation, thus improving precision.

The combination of using various remote cameras can be utilized to support better physical-virtual objects alignment, with a focus on dynamic physical objects. Outdoor environments are dynamic with physical moving objects. In the situation where viewpoints of the user and the remote cameras overlap, it is possible to perform triangulation to track the locations of dynamic physical objects. Dynamic see-through is a method proposed by Barnum and colleagues [22] to visualize physical dynamic objects that are obstructed from the user's viewpoint, using remote cameras. By matching the vanishing point in the remote camera and the user's camera

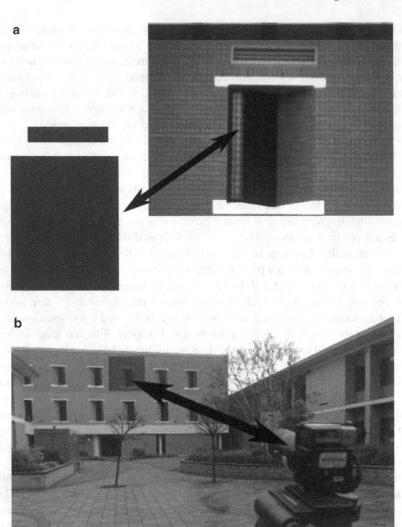

Fig. 4 Combined alignment (**a**) Proper alignment between virtual objects and augmented view-port's camera stream (**b**) Better understanding between the physical camera's location and the physical world. (**c**) A and B leads to more precise alignment between virtual objects and a distant physical building

and applying a 2D image homography between parallel planes, called image homol-ogy, the dynamic see-through method enables tracking and rendering of remote moving objects. Similarly, the augmented Earth map visualization [23] utilizes live camera feeds to track and reconstruct virtual representations of dynamic real world

Fig. 4 (continued)

objects, such as cars and people. The virtual representations are then correctly overlaid on a planar map-view plane of the environment, matching the location of their physical counterparts in the physical environment. We suggest applying similar techniques to the augmented viewport, by utilizing the remote cameras. The positions of dynamic objects can be triangulated from overlapping remote cameras or the user's viewpoint. Once the locations of dynamic objects are known, we can increase the flexibility in aligning virtual objects to dynamic physical objects, to improve precision in object manipulation.

2.5 Input devices

We plan to improve input devices to achieve higher task precision. Considering that even with close hand operations, manipulation tasks are still affected by freehand and sensor errors. We are motivated to design new input devices to support discrete movements for translation, rotation, and scale. The devices will be required to satisfy the guidelines for wearability design [24] such as body placement, appropriate shape, fitting, and most of all, supporting intuitive task execution. One of the most important constraints for input devices for mobile AR system is portability. On the one hand, new input devices should not encumber the user nor hinder with other tasks while not in use. On the other hand, we are interested in the design of new input devices to reduce freehand errors, thus requiring certain tactile feedback and discrete movements. How can a design of new input devices both support a high level of tactile feedback and be uncumbersome and highly wearable?

The augmented viewport technique may benefit from the design of new input devices to support data entry. Desktop-based CAD applications utilize menus, dialogs, or toolbars to allow direct input of exact measurement data from the user. This approach has the highest level of precision; however, it is the least supportive of direct manipulation and contextual visualization.

Text input is an ongoing research problem in the areas of wearable computing and mobile augmented reality systems. Over the years, we have seen solutions developing from physical devices such as belt-mounted mice, forearm keyboards, to virtual onscreen keyboards [25]. Design towards wearability achieves the solution of a glove-based text input mechanism, called the Chording glove [26]. The glove maps combinations of finger presses on one of the hands and a control button press on the other hand to a table of characters. Hand gestures have also been investigated for intuitive textual input. The work by Liu and colleagues [27] presents a technique to allow the user to form hand gestures or perform writing in midair to provide input into the system. Interest in speech recognition for hands-free text input for wearable computers is also prevalent. The various challenges for speech input are distortion and ambient/environmental noises, accuracy, and simultaneous cognitive load [28, 29]. The requirements of specific outdoor AR applications, and the wearabiity and usefulness of the newly designed input devices need to be considered before deciding on the feasibility of new input devices for data entry.

Within the domain of text input for precise manipulation, however, the requirements are more confined. We are only required to support input of digits and common symbols for measurement inputs, instead of the whole alphabet. The Twiddler keyboard, a handheld input device with a 3×4 button matrix, achieves up to 60 words per minute in typing speed [30]. Typing speed is not a requirement for measurement inputs, considering that the input task is only conducted on as-needed basis. It is required, however, to be convenient and effortless, so as not to interfere with the current manipulation task at hand. Therefore, the focus on high wearability, such as comfort of use, as mentioned previously, is more important.

3 Conclusion

We have presented our research position in the problem of precise action at a distance for outdoor AR systems. Based on our technique of the augmented viewport, we have identified current research challenges, including the utilization of various types of remote cameras, collaboration features, better visualization of the cameras' views, precision by snapping, and improved input devices.

References

1. Piekarski W, Thomas BH Tinmith-evo5 - an architecture for supporting mobile augmented reality environments. In: *Proceedings IEEE and ACM International Symposium on Augmented Reality*, 2001. pp 177–178

2. Hoang TN, Thomas B Augmented viewport: An action at a distance technique for outdoor AR using distant and zoom lens cameras. In: *International Symposium on Wearable Computers*, Seoul, South Korea, 2010
3. Kruger W, Bohn C, Frohlich B, Schuth H, Strauss W, Wesche G (2002) The responsive workbench: A virtual work environment. *Computer* 28 (7):42–48
4. Murray N, Goulermas J, Fernando T Visual tracking for a virtual environment. In: *Proceedings of HCI International*, 2003. pp 1198–1202
5. Stoakley R, Conway MJ, Pausch R (1995) Virtual reality on a WIM: Interactive worlds in miniature. *Proceedings of the SIGCHI conference on Human factors in computing systems*:265–272
6. Pausch R, Burnette T, Brockway D, Weiblen ME (1995) Navigation and locomotion in virtual worlds via flight into hand-held miniatures. Paper presented at the *Proceedings of the 22nd annual conference on Computer graphics and interactive techniques*
7. Mine MR, Brooks Jr. FP, Sequin CH (1997) Moving objects in space: Exploiting proprioception in virtual-environment interaction. *Proceedings of the 24th annual conference on Computer graphics and interactive techniques*:19–26
8. Pierce JS, Stearns BC, Pausch R (1999) Voodoo dolls: Seamless interaction at multiple scales in virtual environments. Paper presented at the Proceedings of the 1999 *symposium on Interactive 3D graphics, Atlanta, Georgia, United States*
9. Pierce JS, Forsberg AS, Conway MJ, Hong S, Zeleznik RC, Mine MR (1997) Image plane interaction techniques in 3d immersive environments. *Proceedings of the 1997 symposium on Interactive 3D graphics*
10. Piekarski W, Thomas BH (2004) Augmented reality working planes: A foundation for action and construction at a distance. *Third IEEE and ACM International Symposium on Mixed and Augmented Reality*:162–171
11. Poupyrev I, Ichikawa T, Weghorst S, Billinghurst M Egocentric object manipulation in virtual environments: Empirical evaluation of interaction techniques. In: *Computer Graphics Forum*, 1998. Citeseer, pp 41–52
12. Bowman D, Kruijff E, LaViola J, Poupyrev I (2005) *3d user interfaces - theory and practice*. Addison Wesley, USA
13. Poupyrev I, Billinghurst M, Weghorst S, Ichikawa T (1996) The go-go interaction technique: Non-linear mapping for direct manipulation in VR. *Proceedings of the 9th annual ACM symposium on User interface software and technology*:79–80
14. Bowman DA, Hodges LF (1997) An evaluation of techniques for grabbing and manipulating remote objects in immersive virtual environments. *Proceedings of the 1997 symposium on Interactive 3D graphics*
15. Bowman D, Koller D, Hodges L Travel in immersive virtual environments: An evaluation of viewpoint motion control techniques. In: *IEEE 1997 Virtual Reality Annual International Symposium*, 1997. pp 45–52
16. Schmalstieg D, Schaufler G (1999) Sewing worlds together with seams: A mechanism to construct complex virtual environments. Presence: *Teleoperators & Virtual Environments* 8 (4):449–461
17. Avery B, Piekarski W, Thomas BH (2007) Visualizing occluded physical objects in unfamiliar outdoor augmented reality environments. Paper presented at the *Proceedings of the 6th IEEE and ACM International Symposium on Mixed and Augmented Reality*
18. Wither J, DiVerd S, Hollerer T Using aerial photographs for improved mobile ar annotation. In: *IEEE/ACM International Symposium on Mixed and Augmented Reality*, 22-25 Oct. 2006. pp 159–162
19. Zimmons P, Panter A (2003) The influence of rendering quality on presence and task performance in a virtual environment. Paper presented at the Proceedings of the IEEE Virtual Reality 2003
20. Hirose K, Ogawa T, Kiyokawa K, Takemura H Interactive reconfiguration techniques of reference frame hierarchy in the multi-viewport interface. In: *IEEE Symposium on 3D User Interfaces.*, March 2006. p 73

21. Bier EA (1990) Snap-dragging in three dimensions. SIGGRAPH *Computer Graph* 24 (2):193–204. doi:http://doi.acm.org/10.1145/91394.91446

22. Barnum P, Sheikh Y, Datta A, Kanade T Dynamic seethroughs: Synthesizing hidden views of moving objects. In: *8th IEEE International Symposium on Mixed and Augmented Reality*, 2009. pp 111–114

23. Kim K, Oh S, Lee J, Essa I (2009) Augmenting aerial earth maps with dynamic information. Paper presented at the *Proceedings of the 2009 8th IEEE International Symposium on Mixed and Augmented Reality*

24. Gemperle F, Kasabach C, Stivoric J, Bauer M, Martin R Design for wearability. In: Wearable Computers, 1998. *Digest of Papers. Second International Symposium on, 1998*. p 116

25. Thomas B, Tyerman S, Grimmer K (1998) Evaluation of text input mechanisms for wearable computers. *Virtual Reality* 3 (3):187–199

26. Rosenberg R, Slater M (1999) The chording glove: A glove-based text input device. *IEEE Transactions on Systems, Man, and Cybernetics, Part C: Applications and Reviews*, 29 (2):186–191

27. Liu Y, Liu X, Jia Y Hand-gesture based text input for wearable computers. In: *ICVS '06. IEEE International Conference on Computer Vision Systems*, 2006. pp 8–8

28. Starner TE (2002) The role of speech input in wearable computing. *IEEE Pervasive Computing*, 1 (3):89–93

29. Shneiderman B (2000) The limits of speech recognition. *Commun ACM* 43 (9):63–65

30. Lyons K, Starner T, Plaisted D, Fusia J, Lyons A, Drew A, Looney EW (2004) Twiddler typing: One-handed chording text entry for mobile phones. Paper presented at the *Proceedings of the SIGCHI conference on Human factors in computing systems*, Vienna, Austria

Design Recommendations for Augmented Reality based Training of Maintenance Skills

Sabine Webel, Ulrich Bockholt, Timo Engelke,
Nirit Gavish, and Franco Tecchia

Abstract Training for service technicians in maintenance tasks is a classical application field of Augmented Reality explored by different research groups. Mostly technical aspects (tracking, visualization etc.) have been in focus of this research field. In this chapter we present results of interdisciplinary research based on the fusion of cognitive science, psychology and computer science. We focus on analyzing the improvement of Augmented Reality based training of skills which are relevant for maintenance and assembly tasks. The skills considered in this work comprise sensorimotor skills as well as cognitive skills. Different experiments have been conducted in order to find recommendations for the design of Augmented Reality training systems which overcome problems of existing approaches. The suggestions concern the fields of content visualization and multimodal feedback.

1 Introduction

As the complexity of maintenance and assembly tasks can be enormous, the training of the user to acquire the necessary skills to perform those tasks efficiently is a challenging point. A good guidance of the user through the training task is one of the key features to improve the efficiency of training. From previous research it can be derived, that Augmented Reality (AR) is a powerful technology to support training in particular in the context of industrial service procedures. Instructions on how to assemble/ disassemble a machine can directly be linked to the machines to be operated. Within a cooperation of engineers and perceptual scientists we are working on the

S. Webel (✉)
Fraunhofer Institute for Computer Graphics Research IGD,
Fraunhoferstr. 5, 64283 Darmstadt, Germany
e-mail: sabine.webel@igd.fraunhofer.de

L. Alem and W. Huang (eds.), *Recent Trends of Mobile Collaborative Augmented Reality Systems*, DOI 10.1007/978-1-4419-9845-3_5,
© Springer Science+Business Media, LLC 2011

improvement of AR-based training with the aim to propose design criteria for Augmented Reality based training systems. Here we focus on the training of *procedural skills*. Procedural skills are the ability to follow repeated a set of actions step-by-step in order to achieve a specified goal. It is based on getting a good representation of a task organization: *What* appropriate actions should be done, *when* to do them (appropriate time) and *how* to do them (appropriate method). The procedural skill is the most important skill in industrial maintenance and assembly tasks. Regarding the improvement of training of procedural skills we have analyzed the following points:

Visual Aids: Considering Augmented Reality overlay we can compare two different kinds of visualization: *Direct visual aids*, which means that we are superimposing 3D animations related to a specific task directly onto the corresponding machine parts to be operated; and *indirect visual aids*, which means that we are visualizing annotations related to tracked machine parts which provide the documentation of how to assemble or disassemble this part. An appropriate application of those visual aids may overcome problems of content visualization and may provide a good visual guidance.

Mental Model Building: The so called *mental model* describes the user's internal representation of a task. It has been evaluated that the learner's performance can be improved, when elaborated knowledge is provided to the learner during the task performance [21]. Hence, the visualization of context information may have the potential to support the trainee's mental model building.

Passive Learning Part: A passive learning part, in which the trainee is not active and only receives information and instructions about the task (or device) in general, enables to dedicate part of the training to higher-level instructions. Those higher-level instructions may give the trainee a view of the global picture and not just step-by-step directions. The enhancement of the training system with such a passive learning part may improve training.

Simple Haptic Hints: Haptic hints provide an additional supportive feedback for the user during the training. As they do not prevent the free exploration of the task, they may be helpful for training.

For the examination of these assumptions basic experiments have been performed in order to validate the efficiency of the theses in AR training. In our sample training application a procedure consisting of about 40 more basic assembly steps is considered. One training procedure consists of assembling a linear actuator, another one of disassembling it. We have implemented the different assumptions above and included in the training application. Based on this application cognitive scientists analyzed the impact of the implemented features. At the beginning of each training cycle, the trainer configured the application in terms of enabling or disabling the features. Thus, the features could be considered individually or in combination. The trainee was using a video see-through head-mounted-display (HMD) during the training. On basis of these experiments we propose criteria for the design of Augmented Reality based training systems for maintenance and assembly tasks.

2   Related Work

Until now numerous studies presented the potential of Augmented Reality based training simulators and its use in guidance applications for maintenance tasks. Schwald et al. describe an AR system for training and assistance in the industrial maintenance context [20]. Magnetic and optical tracking techniques are combined to obtain a fast evaluation of the user's position in the whole set-up and a correct projection for the overlays of virtual information in the HMD that the user is wearing. The work discusses the usage of the system, the user equipment, the tracking and the display of virtual information.

In [12] a spatial AR system for industrial CNC-machines, that provides real-time 3D visual feedback by using a holographic element instead of using user worn equipment (like e.g. HMD). To improve the user's understanding of the machine operations, visualizations from process data are overlaid over the tools and workpieces, while the user can still see the real machinery in the workspace.

Reiners et al. introduce an Augmented Reality demonstrator for a doorlock assembly task, that uses CAD data directly taken from the construction/production database to allow the system to be integrated into existing infrastructures [14].

An Augmented Reality application for training and assisting in maintaining equipment is presented in [8]. Using textual annotations and pointing arrows projected into the users view in an HMD, the user's understanding of the basic structure of the maintenance task and object is improved. In [18] Salonen et al. evaluate how an AR system for assembly work for a real setting in a factory should be designed. In this context the advantages and disadvantages of Augmented Reality techniques are explored. Also in [24] the appliance of AR in industrial applications considering the user tracking, interaction and authoring is analyzed and evaluated.

Tümler et al. focus in their work [23] on how long term usage of Augmented Reality technology in industrial applications produces stress and strain for the user. They present user studies comparing strain, measured based on the analysis of Heart Variability, during an AR supported and a non-AR supported work task. Optimal AR systems (no lag, high quality see-through calibration, weight-reduced HMD etc.) are considered as well as non-optimal ones.

A pilot study about the use of wearable Augmented Reality in teaching urban skills training (using the example of room clearing in teams) in order to evaluate the effectiveness of AR for training is presented in [4].

Also in the field of medicine the importance of Augmented Reality based training approaches is growing. Blum et al. [3] propose an AR simulator to improve the training of ultrasound skills. An ultrasound probe is tracked and the corresponding ultrasound images are generated from a CT (computed tomography) volume. Using in-situ visualization of the simulated ultrasound slice and the human anatomy, a deeper understanding of the slice and the spatial layout can be achieved. Furthermore, the trainees performance of a ultrasound task is synchronized to replay of the trainees performance to the experts performance and synchronized replay is shown. An Augmented Reality dental training simulator with a haptic feedback device, to incorporate kinematic feedback and hand guidance, is described in [15]. The simulator provides a extremely realistic conditions for the trainees by combining 3D tooth and tool models upon the real-world view using an HMD.

Other important questions in the context of designing Augmented Reality training systems are how tracking accuracy impacts the applicability of AR in the industry [13], how context information should be provided [16] and how registration errors affect the users perception of superimposed information and the interaction with it ([10, 17]). In [13] the importance of tracking accuracy in industrial Augmented Reality application is elaborated and demonstrated using a sample system. Robertson et al. [17] conducted the effects of registration errors, more precisely 2D position errors, on a desktop assembly application. They compared three cases of registration errors: no error, fixed translation error (constant direction offset of the virtual overlay during the whole assembly task), and random translation error (random magnitude and direction offset of the virtual overlay). In the user trials, a graphical hint that indicates where to place an object was shown to the users subject to one of the three error cases. In half of the trials, also another graphical object that corresponded to a real object was displayed to provide context information for error case currently applied. The trials showed that the users achieved best results in the case of no registration error the worst results in the case of random registration error. Furthermore it has been pointed out, that the contextual information is not really needed when there is a perfect registration, but that it has an enormous positive effect on the user's error rate and performance time of the assembly when registration errors are present.

In another work [16] Robertson et al. evaluated the effects of graphical context in fully registered AR, in non-registered AR, in a Head-Up-Display (HUD) with permanently visible graphics in the field of view and in a HUD with graphics only visible when the user is looking to the side. The study showed, that registered AR led to better performance results than non-registered AR and graphics displayed in the field of view on a HUD.

3   Learning strategies

With the aim of our work we focus on the improvement of the acquisition process of procedural skills. Based on previous research and experiments, the following learning strategies have been defined:

- *Combine direct visual aids and indirect visual aids during the training program*: Apply direct visual aids only at the beginning of the training to help the unskilled trainee to build up his self confidence and to prevent him from performing errors. Use them only if their representation is suitable for the task to perform. Then switch to indirect visual aids, to enhance the user's comprehension of the training task.
- *Use special visual aids as "device display" to enhance the trainee's mental model of the task*: Use a special visual aid as a display of the device being

assembled/disassembled during the training task, including both the condition of the device before the current set of steps and the condition after. This has the potential to support the trainee's mental model building and gives the trainee a view of the whole picture.

- *Add a passive learning part*: Extend the training system by a passive learning part, in which the trainee is not guided step by step through the task, but rather receives more general (i.e. structural) information about the task and the device.
- *Use haptic hints*: Add an additional (haptic) mode of information to the training in order to provide the user abstract haptic hints during training. They do not prevent the free exploration of the task. Thus, the mental model building process can be supported. The hints may be given by using simple devices like a vibrotactile bracelet.

In the following those learning strategies will be analyzed in detail.

3.1   Visual Aids

In the field of Augmented Reality training we can compare two different kinds of visualization: On the one hand we can use *direct visual aids*, where 3D models and animations related to a specific task are augmented directly onto the corresponding real model, which is a common approach. On the other hand, we introduce the use of *indirect visual aids*. In this case, task related annotations are visualized which contain the necessary documentation about what to do and how to do it.

3.1.1   Indirect Visual Aids

For the realization of indirect visual aids in Augmented Reality we adopt the concept of annotations. Such annotations consist of two parts: a pointer, that is attached to a patch in the video image, and a corresponding annotation content, which is displayed on user demand. The pointer overlay provides the spatial information. It highlights the object or the section of the real device, where information for the user is available. The annotation content is visualized screen aligned in a fixed area of the screen (see Figure 1). It can consist of different types of multimedia information, such as text, images, videos and rendered 3D information. Displaying information using such a "pointer-content" metaphor is a very intuitive approach, since it is a typical way to add annotations into documents in the everyday life (e.g. in a text document: the user underlines words in the text block and writes the corresponding explanatory notes at the margin). Hence, the user understands it very easily and learns quite fast how to deal with such kinds of visual aids.

As mentioned in chapter 2, Robertson et al. explored in their studies ([16, 17]) that registration errors have a negative impact on the user's task performance, when no context information visualizing the registration error is given. Since we do not

Fig. 1 Exemplary indirect visual aid: The annotation pointer (highlights the object for which information is available for the user (visualized as pulsatile yellow circle). The corresponding information content is displayed on demand in a fixed area of the screen (right half of the screen)

focus on a special application, but rather on maintenance and assembly tasks in general, the possibility of including such a reference visualization in any use case can not be guaranteed. Also the tracking of objects and environment in real life scenarios is still a challenging field, so a perfect registration of all necessary components during the training can not be assured for arbitrary applications. By using indirect visual aids we can overcome this problem: The pointer acts as a highlight, not as a precise 3D overlay. The superimposition of a 3D animation of the relevant object requires a high tracking accuracy, since an imprecise overlay of the animated 3D model with the real model would interfere the user's recognition and comprehension of the action to perform and would increase his error rate. When using the "pointer-content" metaphor, the annotation content linked to the pointer highlight clarifies definitely which action to perform (cf. Figure 1). Through this combination of tracking depended spatial information and tracking independent content visualization the user can at any time clearly observe the necessary information while never loosing touch with the device.

Apart from all that, indirect visual aids offer the possibility to include a remote component in the training process. It turned out that the use of effective trainer-trainee communication during training can facilitate learning and improve training [7]. A subject matter expert can ameliorate the learning process by enhancing the trainee's knowledge on the best heuristics for the task by giving hints and feedback of results during the performance of the task. In case the expert is not present during the training, but available via a remote connection which allows him for seeing the trainees view (i.e. the camera image) and to transfer digital media, he can create online situation- and individual-related indirect visual aids for the trainee. To do that, he can for instance take a snapshot of the user's view, scribble or draw some information on it and select a patch in the camera image to which the annotation shall be attached. Based on that data an annotation pointer is created at the selected patch, which is linked to the annotated snapshot (annotation content). However, even if the expert or trainer is not locally available, he can nevertheless support the trainee.

During the user training it could be observed, that the concept of indirect visual aids improved the usability of the training system. As the visualization of the annotation content is independent of the tracking and camera movement, it is always clearly visible and legibly; even if the camera is moving very fast (what happens often if it's mounted on a HMD e.g.) and the tracking is disturbed. When we were applying indirect visual aids, the trainees often used the HMD as "reference" after he identified the part to which the aid was linked. More precisely: Once he understood the spatial information (i.e. the object the pointer is highlighting) and the task to perform (the annotation content), he did not care about the tracking anymore. He freely moved his head to feel comfortable. Then he tried to perform the task, seeing through his real eyes without using the HMD (i.e. he looked under the HMD at the real model). To get the information again, he did not reinitialize the tracking, but rather looked up in the HMD to observe the screen-aligned information. Furthermore, even if the registration was not perfect, the user could identify the correct object to deal with. From these results we deduce the following recommendations:

- Indirect visual aids are useful, if a high stability and accuracy of tracking can not be guaranteed.
- Indirect visual aids are helpful, if remote training support shall be provided.

3.1.2 Direct vs. Indirect Visual Aids

If we consider the aspect of content generation, we can notice that the content generation process has always been a critical factor in the development of Augmented Reality based training systems. The generation of 3D content (3D models, 3D animations etc.) usually takes time and requires a huge effort. Shooting a short video or taking a picture of a task or device means much less effort. Often those videos and pictures already exists, as they are utilized in traditional training programs. Consequently, the use of indirect visual aids diminishes the problem of content generation enormously.

When direct and indirect visual aids were compared in the experiments the following points could be observed: Giving the trainee direct visual aids in terms of overlays consisting of animated 3D models caused problems in some use cases and was not suitable for any task or device. The superimposed 3D model could occlude parts of the real model, which were important for the trainee's depth perception. A poor depth perception complicated the trainee's interaction with the device and confused him.

As already mentioned, different tasks require different kinds of media content to provide the best possible support for the trainee. For example this becomes apparent, when the trainee had to perform tasks in which the manner of interacting with the device was a very important factor (e.g. the user needs to remove a plug from a damageable board while fixing the board carefully with one hand in order to avoid to damage it). Presenting the user a short video, that showed exactly what to do and how to interact was easier for him to understand than a complicated 3D animation, and he could perform the task more quickly.

When the trainee needed to execute tasks which require a precise positioning of the overlay (e.g. one of two adjacent plugs on a board shall be removed) or in which small machine parts were included, it was easier for him if indirect visual aids were applied. The pointer, that roughly highlighted the dedicated section, gave the trainee enough spatial information to interpret the annotation information, which showed in detail which plug to remove (e.g. a 3D model of this section containing both plugs, and the one to remove is animated).

Anyhow, it was not always necessary to apply indirect visual aids. In case the task did not require a highly precise superimposition and the task execution itself is not very complex (e.g. the cover of the actuator had to be removed), the use of direct visual aids (e.g. a 3D animation showing the lifting of the cover) provided a fast and easy to understand information. The usage of direct visual aids, which guide the user step by step through the task providing permanently all necessary information, turned out to be suitable in early stages of the training program in order to prevent unskilled trainees from making errors. Based on these observations we suggest, that indirect visual aids are advantageous if:

– direct 3D superimpositions occlude important parts of the device
– the manner of interaction with the device is an important factor for the task
– the task requires a precise superimposition and the a perfect registration can not be provided

Mainly in early stages of the training program the use of direct visual aids, which guide the user step by step through the task providing all the information he needs to know, may be advisable (even though it does not promote the learning process). This strong guidance can prevent unskilled trainees from making errors, because the task representation (i.e. the 3D visualisation) is permanently visible.

3.2   Mental Model Building

It has been explored that the performance of a learner of a procedural skill becomes more accurate, faster, and more flexible when he is provided with elaborated knowledge ([5, 1, 21]). That means, that the learner's performance increases when *how-it-works knowledge* (also called *context procedures*) ([9, 5, 21]) is provided in addition to the *how-to-do-it knowledge* (also called *list procedures*). According to Taatgen et al., when elaborated knowledge is given, the learner is able to extract representations of the system and the task, which are closer to his internal representation, and as a result performance improved [21]. This internal, psychological representation of the device to interact with can be defined as *mental model* [2]. Hence, it is assumed

Fig. 2 The "mental model view" containing the grouped task is visualized in in specified area (top-right), the yellow highlight gives spatial information about the current step to perform

that a display of the device which is assembled/disassembled during the training task has the potential to support the trainee's mental model building.

When the device, or rather the assembly procedure, is complex, it is better to present the user only those sub-parts of it which are relevant for the current step the user has to perform, instead of presenting the entire model ([11, 1]). Furthermore, people think of assemblies as a hierarchy of parts [1], where parts are grouped by different functions (e.g. the legs of a chair). Hence, the hypothesis is that the displayed sub-part of the assembly task should include both the condition of the device before the current step and the condition after. This hypothesis is based on the work of Taatgen et al. [21], in which it is shown that instructions which state pre- and post-conditions yield better performance than instructions which do not. Reviewing this it can be concluded, that the user's mental model building process can be improved by using a context visualization providing context information.

We conclude from this, that during the training session it is necessary to consider the state of knowledge of the potential user. In order to let the user understand the context of the actions to perform we follow the idea of the *mental model*: Successive assembly steps belonging to a logical unit should be grouped and provided as context information. Thus, the complexity of a superimposed assembly step becomes scalable. Furthermore, the grouped tasks should contain the condition of the device before the current assembly task and after.

Figure 2 presents an exemplary grouped task of our training application. Here the main task of removing a cover from the linear actuator is presented through a synthetic animation encapsulating the complete process. At the beginning of the animation the linear actuator is shown in its state before the cover is removed, at the end of the animation it is shown how it should look like after removing the cover. This animation of the "cover remove process" is displayed in a dedicated area on the display, the "mental model view". Its visibility can be toggled on demand. In addition the annotation pointer (yellow highlight) provides the user spatial information about the current step to perform (remove the screw).

Observing the trainees showed the following: When an overview over the sub-process to complete (removing the cover), what is part of the entire process (disassemble the actuator), was presented to the trainee, he got a better understanding of the single steps he needed to perform (e.g. loosen screw 1, loosen screw 2,..., remove cover) and thus he could easier estimate the next step. This accelerated his performance time for a complete training cycle and thus accelerates the complete learning process. From this we conclude that a "mental model visualization" is suitable to

- support the user's mental model building process
- improve the user's performance of the task and to accelerate the learning process

3.3 Passive learning part

The use of visual guidance tools in procedural task training is considered to improve performance [22]. But, simplifying training with visual guidance tools may also harm skill acquisition, because of the possibility of inhibiting active task exploration. To illustrate this we can consider a car driver guided by a car navigation system: this driver typically has less orientation than a car driver orienting with the help of maps and street signs. Active exploration naturally takes place if transferring the information about the task during training is accompanied with some difficulties, forcing the trainee to independently explore the task. When such difficulties are reduced, active exploration may not take place. Strong visual guidance tools impede active exploration because they can guide the trainee in specific movements and thus inhibit the trainee's active exploratory responses [7].

A traditional Augmented Reality training system guides the trainee step-by-step through an assembly task by permanently showing superimposed instructions. Leading the trainee through every single step causes too intensive instruction and thus restrains rather than stimulates the trainees participation. In our application passive learning parts have been included into the training process. Only a structural information about the task is provided to the user permanently. The detailed information about a task is only visualized on demand. Also the context information in terms of the "mental model view" is only displayed if the user induces it and can be completely deactivated by the trainer. A progress bar, which indicates the current state of progress regarding the entire task and the progress inside logical unit (i.e. mental model) is visualized continuously. Our observations showed, that the trainee's performance time of the whole assembly task was improved (i.e. the learning process was enhanced) by including a passive learning part in the training program.

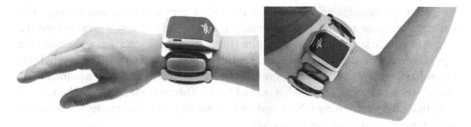

Fig. 3 The vibrotactile bracelet can give the user haptic hints

Also the trainee's comprehension of the next step to perform was accelerated using passive learning elements. Hence, we recommend that passive learning parts should be included in the training program, as they do not impede active exploration and improve the user's performance in training the entire task.

3.4 Haptic hints

The aim of haptic hints is to offer training support that will not deteriorate performance, as they do not prevent the free exploration of the task, and can help in the execution of fine motor trajectories by adding an additional haptic mode of information. Vibrotactile feedback is generated with devices that apply vibration stimuli to the human skin. The use of vibrotactile feedback is versatile and useful to enhance the level of intuitive perception of and interaction with the environment. Those haptic hints do not really guide the user, but rather provide a more subliminal advices to support the user's comprehension of the task to perform. Apart from that, they can communicate a validation of the user's action (i.e. feedback if the user grasped the right tool). This is a significant factor, as it can prevent the user from performing errors at an early stage. Furthermore, such subliminal information give the trainee a more structural information about the task [6], which may enhance the mental model building process.

We provided the haptic hints via a vibrotactile feedback bracelet shown in Figure 3. This device applies vibration stimuli to the human skin. Such a vibrotactile feedback is useful to enhance the level of intuitive perception of the environment and hence the interaction with the environment [19]. The bracelet is equipped with six vibration actuators which are placed at equal distance from each other inside the bracelet (i.e. around the arm). The intensity of each actuator can be controlled individually [19]. Thus, various sensations can be produced. For example, applying repeatedly a short pulse in parallel to all actuators produces a sensation like "knocking"; giving circular impulses around the arm leads to a sensation that prompts the user to "rotate the arm" (what corresponds to circular motions like loosening a screw). A "warning" sensation (e.g. the user took the wrong tool) can be produced by giving one stronger and longer impulse to all actuators in parallel. Anyway, via the bracelet the user is getting vibrotactile information about the task he has to perform and if he is performing it right or wrong.

Applying rotational haptic hints at the user's arm when he was supposed to perform a circular motion (like loosening a screw) helped him to understand the task. After the first application of the hint (to tell him to loosen a screw) he was able to perform similar steps without asking for the annotation content. The haptic hints in combination with the spatial information provided by the annotation pointer were enough for him to recognize, what he had to do. Through warning hints he noticed early that he is performing falsely and that he has to check his actions. Considering these aspects we deduce the following:

- Haptic hints can be provided to the user by using simple and more abstract devices (like vibrotactile feedback devices)
- Haptic hints are useful to improve the learning process and to prevent the user from performing errors at an early stage

4   Results and Conclusion

Based on previous research and experiments performed in cooperation with human factors scientists recommendations for the design of Augmented Reality based training systems have been established.

Indirect visual aids linking instructions as annotations to tracked machine parts shall be included in the training system, since they are often more efficient than direct visual overlays. The use of indirect visual aids allows us to provide multimedial information (e.g. images, videos, text, 3D content) about the task to perform for the trainee, instead of showing him only an animated 3D model overlaid with the camera image. Moreover, applying those indirect aids promotes the trainee's learning process, as they do not impede the user's active exploration of the task. This approach requires less tracking accuracy than direct superimpositions of animated 3D models, since only a section on the real model framing the part to interact with is highlighted. The concrete task is shown in the annotation content, which is displayed on demand of the trainee. As visualization of the annotation content is tracking independent, it can be clearly observed even if the tracking is interrupted. Besides this, indirect visual aids form a passive learning element, what improves the learning process. Apart from that, this technique eases the content generation process and opens new opportunities for distributed Augmented Reality applications as images, videos or scribblings can be generated online by a remote expert and transferred to the trainee or service technician.

At early stages of the training program we recommend to apply direct visual aids, as they help unskilled trainees to build up self confidence and prevent them from making errors.

We also propose to provide elaborated knowledge during the training to support the user's mental model building process. Furthermore, we suggest to include passive learning parts and haptic hints in the training procedure, as they enhance the user's comprehension and internalization of the training task, and thus improve training.

Following these recommendations, critical problems of traditional Augmented Reality based training systems, such as tracking inaccuracies and adequate information visualization, can be overcome.

5 Future Work

In cooperation with human factor scientists we will further evaluate our observations described in this chapter by more detailed experiments. User studies will be conducted to confirm the results. Furthermore we plan to examine the optimal information visualization for different devices (e.g. HMD, Ultra-Mobile PCs).

Acknowledgements The presented results have been developed within the context of the European project SKILLS, that deals with the exploration of multimodal interfaces for capturing and transfer of human skills. We thank TECHNION (Israel Institute of Technology) for their experiments and contributions about important cognitive aspects. Furthermore, we thank DLR (German Aerospace Center) for developing and providing the vibrotactile bracelet and LABEIN Tecnalia and PERCRO (Perceptual Robotics Laboratory) for pushing the idea of this work and their contributions.

References

1. M. Agrawala, D. Phan, J. Heiser, J. Haymaker, J. Klingner, P. Hanrahan, and B. Tversky. Designing effective step-by-step assembly instructions. *ACM Trans. Graph.*, 22(3):828–837, 2003.
2. A. Antolí, J. F. Quesada, F. D. Psicología, J. J. Cañas, and J. J. Cañas. The role of working memory on measuring mental models of physical systems. In *Psicológica*, volume 22, pages 25–42, 2001.
3. T. Blum, S. M. Heining, O. Kutter, and N. Navab. Advanced training methods using an augmented reality ultrasound simulator. In *ISMAR '09: Proceedings of the 2009 8th IEEE International Symposium on Mixed and Augmented Reality*, pages 177–178, Washington, DC, USA, 2009. IEEE Computer Society.
4. D. G. Brown, J. T. Coyne, and R. Stripling. Augmented reality for urban skills training. In *VR '06: Proceedings of the IEEE conference on Virtual Reality*, pages 249–252, Washington, DC, USA, 2006. IEEE Computer Society.
5. R. M. Fein, G. M. Olson, and J. S. Olson. A mental model can help with learning to operate a complex device. In *CHI '93: INTERACT '93 and CHI '93 conference companion on Human factors in computing systems*, pages 157–158, New York, NY, USA, 1993. ACM.
6. D. Feygin, M. Keehner, and F. Tendick. Haptic guidance: Experimental evaluation of a haptic training method for a perceptual motor skill. In *HAPTICS '02: Proceedings of the 10th Symposium on Haptic Interfaces for Virtual Environment and Teleoperator Systems*, page 40, Washington, DC, USA, 2002. IEEE Computer Society.
7. N. Gavish and E. Yechiam. The disadvantageous but appealing use of visual guidance in procedural skills training. *Proceedings of the Applied Human Factors and Ergonomics Conference (AHFE) '10*, 2010.
8. C. Ke, B. Kang, D. Chen, and X. Li. An augmented reality-based application for equipment maintenance. In *ACII*, pages 836–841, 2005.
9. D. Kieras. What mental model should be taught: Choosing instructional content for complex engineered systems. In *Intelligent tutoring systems: Lessons learned, Lawrence Erlbaum*, pages 85–111. Psotka, J., Massey, L.D., Mutter, S. (Eds.), 1988.

10. M. A. Livingston and Z. Ai. The effect of registration error on tracking distant augmented objects. In *ISMAR '08: Proceedings of the 7th IEEE/ACM International Symposium on Mixed and Augmented Reality*, pages 77–86, Washington, DC, USA, 2008. IEEE Computer Society.
11. L. Novick and D. Morse. Folding a fish, making a mushroom: The role of diagrams in executing assembly procedures. *Memory & Cognition*, 28(7):1242–1256, 2000.
12. A. Olwal, J. Gustafsson, and C. Lindfors. Spatial augmented reality on industrial CNC-machines. In *Society of Photo-Optical Instrumentation Engineers (SPIE) Conference Series*, volume 6804 of *Society of Photo-Optical Instrumentation Engineers (SPIE) Conference Series*, Feb. 2008.
13. K. Pentenrieder and P. Meier. The need for accuracy statements in industrial Augmented Reality applications. In *Proc. ISMAR Workshop: Industrial Augmented Reality*, Santa Barbara, CA, USA, October 2006.
14. D. Reiners, D. Stricker, G. Klinker, and S. Müller. Augmented reality for construction tasks: Doorlock assembly. In *Proc. IEEE and ACM IWAR98 (1st Int. Workshop on Augmented Reality*, pages 31–46. AK Peters, 1998.
15. P. Rhienmora, K. Gajananan, P. Haddawy, S. Suebnukarn, M. N. Dailey, E. Supataratarn, and P. Shrestha. Haptic augmented reality dental trainer with automatic performance assessment. In *IUI '10: Proceeding of the 14th international conference on Intelligent user interfaces*, pages 425–426, New York, NY, USA, 2010. ACM.
16. C. Robertson, B. MacIntyre, and B. Walker. An evaluation of graphical context when the graphics are outside of the task area. In *Mixed and Augmented Reality, 2008. ISMAR 2008. 7th IEEE/ACM International Symposium on*, pages 73–76, 15-18 2008.
17. C. M. Robertson, B. MacIntyre, and B. N. Walker. An evaluation of graphical context as a means for ameliorating the effects of registration error. *IEEE Trans. Vis. Comput. Graph.*, 15(2):179–192, 2009.
18. T. Salonen, J. Sääski, M. Hakkarainen, T. Kannetis, M. Perakakis, S. Siltanen, A. Potamianos, O. Korkalo, and C. Woodward. Demonstration of assembly work using augmented reality. In *CIVR '07: Proceedings of the 6th ACM international conference on Image and video retrieval*, pages 120–123, New York, NY, USA, 2007. ACM.
19. S. Schätzle, T. Ende, T. Wüsthoff, and C. Preusche. VibroTac: An ergonomic and versatile usable vibrotactile feedback device. In *RO-MAN '10: 19th IEEE International Symposium on Robot and Human Interactive Communication*, Principe di Piemonte - Viareggio, Italy, 2010. IEEE Computer Society.
20. B. Schwald, B. D. Laval, T. O. Sa, and R. Guynemer. An augmented reality system for training and assistance to maintenance in the industrial context. In *The 11th International Conference in Central Europe on Computer Graphics, Visualization and Computer Vision 03, Plzen, Czech Republic*, pages 425–432, 2003.
21. N. A. Taatgen, D. Huss, D. Dickison, and J. R. Anderson. The acquisition of robust and flexible cognitive skills. *Journal of Experimental Psychology: General*, 137(3):548–565, 2008.
22. S. B. Thrun. Efficient exploration in reinforcement learning. Technical report, 1992.
23. J. Tümler, F. Doil, R. Mecke, G. Paul, M. Schenk, E. A. Pfister, A. Huckauf, I. Bockelmann, and A. Roggentin. Mobile augmented reality in industrial applications: Approaches for solution of user-related issues. In *ISMAR '08: Proceedings of the 7th IEEE/ACM International Symposium on Mixed and Augmented Reality*, pages 87–90, Washington, DC, USA, 2008. IEEE Computer Society.
24. J. Weidenhausen, C. Knoepfle, and S. D. Lessons learned on the way to industrial augmented reality applications, a retrospective on arvika. *Computers and Graphics*, 27:887–891(5), 2003.

Collaborative Content Generation Architectures for the Mobile Augmented Reality Environment

Daniel Gallego Vico, Iván Martínez Toro,
and Joaquín Salvachúa Rodríguez

Abstract The increasing adoption of smartphones by the society has created a new research area in mobile collaboration. This new domain offers an interesting set of possibilities due to the introduction of augmented reality techniques, which provide an enhanced collaboration experience. As this area is relatively immature, there is a lack of conceptualization, and for this reason, this chapter proposes a new taxonomy called Collaborative Content Generation Pyramid that classifies the current and future mobile collaborative AR applications in three different levels: Isolated, Social and Live. This classification is based on the architectures related to each level, taking into account the way the AR content is generated and how the collaboration is carried out. Therefore, the principal objective of this definition is to clarify terminology issues and to provide a framework for classifying new researches across this environment.

1 Introduction

During the recent years the mobile world has experienced an extremely fast evolution not only in terms of adoption, but also talking about technology. Smartphones have become a reality allowing users to be permanently connected to the Internet and to experience new ways of communication.

Handheld devices are nowadays able to support intensive resource demanding applications, what makes possible for developers to create applications that change the way end users experience the world and communicate with each other. This kind of

D.G. Vico (✉)
Departamento de Ingeniería de Sistemas Telemáticos, Escuela Técnica
Superior de Ingenieros de Telecomunicación, Universidad Politécnica de Madrid,
Avenida Complutense 30, "Ciudad Universitaria", 28040, Madrid, Spain
e-mail: dgallego@dit.upm.es

L. Alem and W. Huang (eds.), *Recent Trends of Mobile Collaborative
Augmented Reality Systems*, DOI 10.1007/978-1-4419-9845-3_6,
© Springer Science+Business Media, LLC 2011

applications can be created using the power of context sensors that are present in this new series of phones. Another important characteristic of this innovative scenario is the integration of applications with social networks that enables mobile collaboration.

Developers have recognized in augmented reality (AR) technology combined with smartphones' power, a rich environment that offers a wide range of exciting possibilities. One of the best is the use of mobile AR to improve collaboration among users.

Mobile Collaborative AR is today an active field of research since applications using these concepts are being deployed every day. The most accepted applications are the ones running on mobile phones compared to other possibilities that are still in a research progress, such as wearable AR.

Within the mobile collaborative AR environment as it is a relatively young technology, there exists a lack of conceptualization. Taxonomies that would allow standardization processes are required. They would be also useful to optimize the research and development processes.

The way of generating and sharing the content is a concrete attribute that differs from one mobile collaborative AR application to another. It can be found that some applications just provide information overlays to users who cannot add any feedback or extra content. Some other applications let the end users add simple or complex content in order to enrich the entire system, even connecting with social networks. There also exists another type of applications based on real time content generation and sharing. Studying these differences in content generation and analyzing directly related parameters such as the architecture and technology needed by each type of application, and the user perceived impact, we have defined a new taxonomy. The taxonomy we present in this chapter is called Collaborative Content Generation Pyramid. It consists of three different levels in which the applications can be classified: Isolated, Social and Live. Each level is technologically and architecturally supported by the lower levels.

We show the advantage of this new taxonomy by using it in order to analyze and classify the currently deployed and well known applications.

The structure of the chapter is as follows: the next section introduces the background of our research, explaining the current context in which this work is based. We then move on to define the proposed Collaborative Content Generation Pyramid, in section 3. We explain next in section 4, how it can be used to classify the current environment, also presenting some interesting ideas of potential Live Level developments. Finally, the last section presents the conclusions revealed after the research was carried out and summarizes the ideas for future work.

2   Background

Before starting with the description of the Collaborative Content Generation Pyramid concept and the different architectures related to it, we are going to provide an overview of the context behind our research in the mobile and AR areas, regarding to how both of them are contributing nowadays to improve the collaborative experience in the mobile world.

2.1 A New Age in the Mobile World

When talking about the mobile world, we notice that it has been evolving very fast in the recent years due to the evolution of mobile devices.

Since the appearance of smartphones, more specifically the Apple iPhone in 2007, the mobile market has drastically changed. Consequently, all the mobile companies are trying to imitate it, therefore it is not wrong to say that the iPhone was a disruptive change in the mobile devices world.

Due to this evolution, a set of technical characteristics have been assumed in order to design mobile devices. Probably the most important thing to take into account is that now the telecommunication companies focus the design considering the user first [1], instead of the technology itself.

A good example of this assertion is the use of multi-touch, which has completely revolutionized the mobile user experience. Perhaps the most important characteristic of this new series of mobile devices is their integration with social networks, in order to be able to provide mobile collaboration by interconnecting the mobile and the Internet worlds via the Social Web, as Reynolds said in [2].

Furthermore, it is important to remark that the power of the current mobile devices is in some cases comparable with the one of a computer, and if we also take into account the context devices they have (e.g. camera, GPS, compass, accelerometer...), there exist a lot of possibilities to enable collaboration in several ways.

2.2 The Rebirth of Augmented Reality

Related to the previous section, we can talk about a recent innovation that has appeared in that context: the use of AR in mobile applications. An AR system, according to Azuma et al [3], must have the following properties:

- Combines real and virtual objects in real environments.
- Runs interactively and in real time.
- Registers real and virtual objects with each other.

As Vaughan-Nichols said in [4], although it is nowadays a hot topic, AR is in fact an old technology.

The first appearance of the "Augmented Reality" concept is attributed to Tom Caudell while he was working for Boing in 1990. However there existed systems that achieved the properties mentioned before ever since 1960s and there are several areas where AR found many applications during the recent years, as we can see in [5] and [3]. It was in 1994 when the idea of AR was perfectly established due to the

Milgram`s Reality-Virtuality continuum [6]. He defined a continuum of real to virtual environment, in which AR takes one part in the general area of mixed reality. Throughout the years, the other parts of the continuum (augmented virtuality and virtual environments), have not reached much importance in the mobile world, and for this reason we are not going to talk about them in this chapter.

Returning to AR, we can say that it has changed the common mobile applications inputs, replacing them with new ones: registration and image acquisition, physical or virtual location, data from the compass and the accelerometer or user's touches. Consequently, the way a developer works has also changed, because he or she must take into account these new kinds of inputs in order to be able to achieve a complete AR application.

Besides this, the way users interact with these augmented systems has changed radically from the traditional mobile applications, because in some manner they interact directly with the real world throughout the application, having a lot of possibilities to live an innovative user experience.

2.3   Mobile Collaborative Augmented Reality

As introduced before, the use of AR in mobile applications is a consequence derived from the set of possibilities that these kind of devices offer to developers. If also we take into account the importance that the collaboration has achieved nowadays in the mobile world due to the penetration of social networks like Facebook or Twitter, the joint of both areas (mobile collaboration and AR) is a direct outcome. This is proved taking into account that several of the 10 disruptive technologies from 2008 to 2012 proposed in [7] are implicated in the mobile collaborative AR.

In the recent years, there have been developed some significant examples in this area with different results. On the one hand, considering wearable mobile devices, Reitmayr and Schmalstieg [8] designed a system capable of providing 3D workspaces for collaborative augmented reality environments. In this way, Mistry et al. [9] defined a wearable gestural interface system capable of showing information into the tangible world (in some cases obtained from the Internet), due to the use of a tiny projector and a camera. Also there are other examples like the project developed by Hoang et al. [10], whose main objective is to connect users through Web 2.0 social networks in a contextually aware manner based on wearable AR technologies; or the research described in the chapter 4 of this book by Hoang and Thomas [11] related to AR wearable computers in an outdoor setting, taking into account the requirement of mobility.

On the other hand, there exist examples of non wearable systems such as the face to face application developed by Henrysson et al. [12] called AR Tennis, which is based on fiducial markers to carry out the tracking. Or the SmARt World framework illustrated in the chapter 3 of this book by Yew et al. [13], designed to enable a smart space where real and virtual objects co-exist in an AR way, allowing users to interact collaboratively with this objects in the smart space using only a smartphone.

Nevertheless, in this chapter we are going to focus on mobile collaborative AR systems that track without using wearable technologies or fiducial markers, because the great amount of AR applications nowadays are supported by mobile phones that normally do not use these tracking techniques. Furthermore, it is important to point out the existence of requirements established by the society for the wearable AR systems to be sufficiently attractive, comfortable, optically transparent and inexpensive, in order to be used in everyday collaboration by users, as Feiner said in [14].

In relation to this, we think that mobile phones are now the best devices to promote collaborative mobile AR applications, because of their world wide acceptance, and more specifically the smartphones are achieving a high affirmation that is going to increase in the next years as we can see in [15].

3 Collaborative Content Generation Pyramid

As AR is already an active and dynamic research and development area, and due to its immaturity and rapid growth, there still exists a lack of conceptualization. As a consequence, some interesting properties like the content generation differences between applications are not placed in any existing taxonomy. Definitions and taxonomies are essential tools that lead to a better understanding and optimization of research and development processes. Nowadays, a big effort is being made to achieve this kind of tools, as can be seen in the work of the World Wide Web Consortium (W3C) to define standards for managing common AR resources including geolocation [16] and camera [17].

While studying the AR environment and the coexisting applications, we have noticed that there is a concrete characteristic that differs from one application to another. This characteristic is the way that the content to be shown as information overlays is generated; the collaboration capabilities each application enables is directly related to this characteristic. According to the idea of different content generation approaches, there are some applications that use a centralized content creation and management approach. Others present information generated by the end user in a real time. Between these counter approaches there is still a wide range of others, each of those requiring a different kind of technology and architecture.

To overcome the lack of a general taxonomy that represents the previously discussed characteristic, we propose a new categorization: the Collaborative Content Generation Pyramid (illustrated by Fig. 1). This taxonomy classifies the AR environment starting from a main criterion which is the content generation as introduced before, but taking into account some other strongly related concepts such as the technology and architectures needed and the user perceived impact. Following these ideas we have defined three different levels (Isolated, Social and Live), which will be explained in details in the following sections. The pyramidal structure of the classification comes from the fact that each level is supported by the architecture and technology of the lower levels.

Fig. 1 Collaborative Content
Generation Pyramid

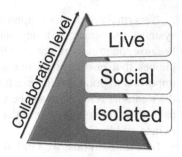

In the next subsections this classification is presented by describing its levels and their characteristics. In section 4, the use of this classification as a categorization tool is illustrated.

3.1 Isolated Level

This level includes any application that uses a centralized content generation and management. That is, every application in which the information forming different overlays is created or gathered by the application team and managed only by the application's servers. The user receives a content that is related to the point of interest (POI) that he is looking at, only if this point is registered with associated content in the application's server. There is no contribution made by the end user to the content repository as the user is just a consumer of information and does not participate in the generation nor improvement of the content. As a result, the collaboration in this level is minimal.

However, this is an important level to study not only because of the existence of applications matching with its characteristics, but also because it is the technological and architectural base for the upper levels.

The architecture that supports this kind of systems is in most of the cases similar to the one showed in Fig. 2. The main part of the scheme is the application server, which contains all the content to be shown as information overlays. This level is named isolated because this server is not related with other services, appearing as the only source of information for the client part. The client part may be a smartphone containing different physical context devices such as camera, compass and accelerometer, being able to access external context devices such as the GPS. Diverse final applications use different sets of these devices to achieve AR.

The applications of this level are mainly based on context awareness. They use the environmental information around the user in order to enhance the mechanisms of AR, improve their performance [18], and be aware of this environment and react accordingly to the user's context [19].

Fig. 2 Isolated Level
architecture

3.2   Social Level

The Social Level is located in the middle of the previously defined pyramid and it refers to any application that presents information layers generated by the collection of different content sources. Not only corporations such as Wikipedia, Ebay, Amazon or the application itself share their data, but also the end users generate and share contents via social networks (Facebook, Twitter, Blogger, etc) or by direct upload to the application server. The user is able to select a POI and attach some information to it that will be stored and shown to other users. The social environment of each user is collaborating dynamically in order to create content that leads to context generation.

In this case, the architecture becomes more complex due to the fact that it has to support the collaboration among many participants, as shown in Fig. 3. The application server must allow dynamic content addition. The client part is similar to the one described in the Isolated Level, with the extra capability of uploading content to the server and accepting it from diverse sources.

From the entire explanation, the mobile collaborative AR starts in this level. The power of social networks combined with the potential of AR and the characteristics of mobile devices, enables the creation of real collaborative applications in which

Fig. 3 Social Level
architecture

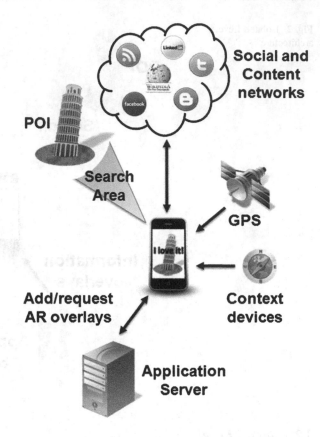

the end user is the one that creates the most interesting information. Other users will consume it as overlays over the real world. Social awareness is generated in this level, what can be used to enable many different types of collaborative applications [20]. The majority of applications being developed and released nowadays takes place in this level of the pyramid and since their importance and acceptance are growing, new ways of collaboration are to be opened.

3.3 Live Level

The top level within the Collaborative Content Generation Pyramid corresponds to the applications that share information overlays generated in a real time. More explicitly, in these applications two or more users connect with each other and contribute in the live generation of AR content as layers over the POI captured by one of the users. The generated content can be stored in order to be available for future use, or it can be simply session persistent. This level also takes advantage of the social networks as explained in the Collaborative Level.

Fig. 4 Live Level architecture

To support such kind of applications, a completely distributed architecture is needed. Therefore there is more than one possible solution, including P2P. In Fig. 4, an abstraction of a possible architecture is represented, showing the interaction among users, the generation of content over the search area of one of the users, and the sharing of the content. Applications based on this level require all the technology offered by the lower levels and additionally a technological solution for real time communication.

Focusing on collaboration and new ways of interaction among users, there is still a lot to explore in this level. Real time collaboration using AR appears to be in its early stages of development and there is a lack of working applications supporting it, but the opportunities are endless as will be discussed in following chapters.

3.4 Analyzing the Pyramid Characteristics

Depending on a certain level of the pyramid, different characteristics can be identified apart from the ones included in each category definition introduced above. Therefore, applications that are placed in different levels of the classification have common attributes.

One of the attributes studied has to do with the resources needed to support the application that vary from one level to another. This concept can be put as a

Fig. 5 Resources vs Impact comparative

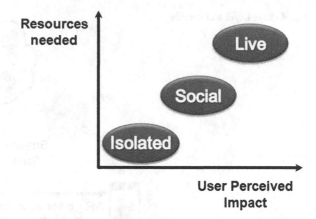

combination of the technological and architectural requirements, where by technological requirements we understand the complexity of the client and server parts (from physical context retrieval devices to fast processing capabilities). When studying the resources needed for each level, it is easy to determine that applications categorized as mainly isolated require less resources. They are based on a centralized server demanding a simpler architecture and network capabilities and they do not allow users to publish their own content, simplifying the client part of the application. The applications classified in the Collaborative Level need more complex server and client parts to allow different sources to generate and include content. This makes an exhaustive use of the network and demands a well defined and in most cases, distributed architecture.

Those applications classified in the Live Level, as any other real time application, need the most of the network to enable live content generation and exchange among different users. Its architecture must be completely distributed, which implies a high complexity, thus each client needs some extra functionalities (e.g. P2P support).

Another characteristic that can be considered is the impact perceived by end users. Starting from the bottom end of the pyramid, the Isolated Level, its users usually perceive a limited impact. Once a user is used to the AR experience an application using a static content source makes them perceive it as an ordinary guide, with no impression of a significant extra value, as it does not allow any collaboration.

As we follow the pyramid, the user perceived impact grows with the collaboration characteristic. According to the social media success, allowing collaboration gives extra value to users' interactions and experiences. At the top end of the categorization, the Live Level applications make users perceive the highest impact as they let them observe and participate in a real time in the content generation.

The comparison between the resources needed and the user perceived impact, following the statements introduced above, is represented in Fig. 5.

4   Classifying Present Applications

After defining the Collaborative Content Generation Pyramid, it is time to use it to evaluate and classify the current environment in AR applications.

We have analyzed a set of applications to present a general view of the current situation. Following the ideas explained in the early sections we take into account only mobile applications, leaving apart systems like wearable AR. Furthermore we have focused on the real market avoiding work-in-progress researches; therefore we have selected the most important applications mainly from the Android Market and the AppleStore. It is necessary to remark that nowadays no deployed applications fit into the Live Level of the pyramid. However we present some possible future developments that could take advantage of this top level collaboration.

This section illustrates how this new classification can be used to classify existing applications. It also shows how it could guide a developer to place his application in the proper level in order to find out the resources needed as well as the density of competitors existing in that level.

4.1   Isolated Level

If we focus our attention in the applications located at the first level of our taxonomy, the Isolated Level, we quickly discover that these kinds of systems were the initial AR applications developed years ago for the mobile markets, because they are quite simple compared to the current applications.

We have selected two examples corresponding to that level. First, the Nearest Tube [21] developed by Acrossair to provide an AR map to inform people about where the nearest tube station is. As we can see, this application only has content generation in one direction: from the application's servers to the smartphone user client.

Another example is Theodolite [22], which offers information about position, altitude, bearing, and horizontal/vertical inclination of the user's smartphone using the context devices and the GPS to consult these data to the application's servers. As we can see, both (Nearest Tube and Theodolite), correspond to the Isolated Level architecture illustrated by Fig. 2, since they are based on context awareness techniques and also, they only use the application's servers to get the AR overlays.

4.2   Social Level

The second level of the pyramid is populated by the latest applications developed. These are the most representative existing examples of how the AR can be used to enable collaboration.

In this case, while studying the social applications available in the different markets, we have found that each application enables a different level of collaboration. Therefore, it is possible to divide this level in two sublevels, low social and high social.

Some applications such as Layar [23] and Junaio [24] belong to the lower part of the Social Level due to the fact that they let users to add simple pieces of information (comments, ratings, etc.) to existing POIs, but the creation of POIs is only achievable for developers. Becoming a developer for this kind of applications requires programming skills, so the average user is not able to generate complete content.

There is another type of applications like Wikitude [25], WhereMark [26] or Sekai Camera [27] that fit in the higher part of the Social Level. The reason is that by using these applications any end user is able to create POIs and to attach simple information that will be visible for the rest of the community.

4.3 Live Level

As we remarked in the subsection 3.3, the Live Level is a research area with a motivating future. The reason for this is that nowadays we cannot find commercial applications that are capable of interconnecting several users to generate AR content in a real time and in a collaborative way, following the architecture illustrated by Fig. 4.

For this reason, now we are going to set out some possible use cases interesting for different areas of knowledge that will have the live properties we described before.

4.3.1 Entertainment

Building applications for this area is probably the best way to achieve high use of the Live Level. In order to establish this level architecture in the mobile world, a lot of small games for the different mobile markets could be quickly developed. Of course more complex games similar to, for example, World of Warcraft, would be killer applications in this area due to the massive users registered and the possibility of being integrated in social networks like Facebook.

4.3.2 Education

We can think about a lot of excellent use cases in this context when we talk about such kind of applications. A lection could be given following a 1 to N user connection model, where there would be one teacher and N students receiving the teachings through the screen of their mobile devices. In this way, interacting with the group by creating AR contents that everybody could see in a real time collaborative experience could be possible. Additionally, if we take a step forward, the application could record the classes in order to reproduce them in the future. Also it exists the possibility of adding AR notes and share the videos with the rest of the class in a social way via Facebook, Twitter, and so on.

4.3.3 Medicine

If we think in an isolated geographical zone (like a small town or village) where it is difficult to access and, as a consequence, receiving medical assistance is complicated, the use of live applications would be quite useful. Thus, we could use an application following the Live architecture to connect directly to the family doctor to for example, show him a person's wound and then ask him how to clean it.

Another use in this context is focused on medical distance examinations, because using the camera we could show a sick person to the doctor, and then he could explain us how to treat the illness via generating graphical or textual information overlays in a real time. Furthermore, if we think in an uncommon illness, with this kind of application, we could interconnect a group of experts from different medicine areas to make a medical examination in a collaborative way.

5 Conclusion and Future Works

Along this chapter we have proposed some important definitions to conceptualize and classify the current paradigm in collaborative mobile AR applications, in order to clarify what are the architectures behind these systems.

We started with an analysis of the current background related to this field, discovering that the joint of both worlds (mobile collaboration and AR) has created a new area where everyday new mobile applications emerge.

Then, we defined the Collaborative Content Generation Pyramid, a taxonomy divided in three levels according to the way the AR content is generated in these mobile applications. Also, the architectures used to build this kind of systems are described to help developers understand what type of application they want to achieve, and what the necessary components are.

After that, we used the taxonomy to classify a set of real applications present in the different mobile markets. Besides, we proposed some use cases of Live Level applications in order to remark that this level is nowadays an interesting research area.

Furthermore, we believe that one of the most important problems of the taxonomy's top levels, i.e., the real time graphic rendering of all the AR content overlays in the client side, will be solved by initiatives like OnLive [28], which pretends to carry out all the graphic rendering related to a gaming platform using cloud computing techniques. Therefore, with this kind of methods, the generation of the AR content for the mobile devices in a cloud computing way could help to reduce the charge on the client side.

Following the idea of using cloud computing techniques in this area, we propose to use cloud-based content and computing power resources to improve the scene, the rendering and the interaction as Luo said in [29], because it is an important factor to increase the power of mobile collaborative AR applications.

Additionally, taking into account that current AR applications have different models to identify POIs, we think that it is important to create standard APIs or communication formats, in order to be able to share overlays created by each user in any application domain, instead of repeating the same AR content in every application related to the same POIs. With this is mind, we have been following the outcomes from the W3C Workshop: Augmented Reality on the Web, where some related research work was proposed. One of it was the suggestion of Reynolds et al. [30] to exploit Linked Open Data for mobile augmented reality applications in order to solve the previous problem of content redundancy.

One more interesting research line could be the addition of another dimension to the taxonomy described along this chapter. The purpose of this new dimension would be to analyze and illustrate the social impact of using this kind of applications, i.e., the number of users collaborating at the same time and how this affects the user experience.

Finally, to conclude, we believe that the definitions presented in this chapter can help to order and classify the current confuse AR environment, and also, they are solid foundations capable of supporting all the present existing AR applications, and all the future and exciting ideas that certainly will appear in this area.

References

1. J. Miller. The User Experience.In *IEEE Internet Computing*, volume 9, no. 5, pp. 90–92. September-October, 2005.
2. F. Reynolds. Web 2.0-In Your Hand. In *IEEE Pervasive Computing*, volume 8, no. 1, pp. 86-88. January, 2009.
3. R. Azuma, Y. Baillot, R. Behringer, S. Feiner, S. Julier and B. MacIntyre. Recent advances in Augmented Reality. In *IEEE Computer Graphics and Applications*, volume 21, no. 6, pp. 34–47. November/December 2001.
4. S. J. Vaughan-Nichols. Augmented Reality: No Longer a Novelty? In *Computer*, volume 42, no. 12, pp. 19–22. December, 2009.
5. R. Azuma. A Survey of Augmented Reality. In *Presence: Teleoperators and Virtual Environments*, volume 6, no. 4, pp.355–385, August 1997.
6. P. Milgram, H. Takemura, A. Utsumi and F. Kishino. Augmented Reality: A class of displays on the reality-virtuality continuum. In *Telemanipulator and Telepresence Technologies*, volume 2351, pp. 282–292. 1994.
7. Gartner Emerging Trends and Technologies Roadshow. *Gartner Identifies Top Ten Disruptive Technologies for 2008 to 2012.* 2008. [Online]. Available: http://www.gartner.com/it/page. jsp?id=681107 [Accessed: August 17, 2010].
8. G. Reitmayr and D. Schmalstieg. Mobile collaborative augmented reality. *Proceedings of ISAR '01* (New York, NY, USA, October 29–30, 2001), pp. 114–123, 2001.
9. P. Mistry, P. Maes and L. Chang. WUW - Wear Ur World - A Wearable Gestural Interface. *Proceedings of Human factors in Computing Systems* (Boston, MA, USA, April 4–9, 2009), ACM, pp. 4111–4116, 2009.

10. T. N. Hoang, S. R. Porter, B. Close and B. H. Thomas. Web 2.0 Meets Wearable Augmented Reality. *Proceedings of ISCW '09* (Linz, Austria, September 4–7, 2009), IEEE Computer Society, pp. 151–152, 2009.

11. T. N. Hoang and B. H. Thomas. Augmented Viewport: Towards precise manipulation at a distance for outdoor augmented reality wearable computers. In *Mobile Collaborative Augmented Reality Systems*, chapter 4, Springer Book, 2010.

12. A. Henrysson, M. Billinghurst and M. Ollila. Face to face collaborative AR on mobile phones. *Proceedings of ISMAR '05* (Vienna, Austria, October 5–8, 2005), pp. 80–89, 2005.

13. A.W.W. Yew, S.K. Ong and A.Y.C. Nee. User-Friendly Mobile Ubiquitous Augmented Reality Computing. In *Mobile Collaborative Augmented Reality Systems*, chapter 3, Springer Book, 2010.

14. S.K. Feiner. The importance of being mobile: some social consequences of wearable augmented reality systems. *Proceedings of IWAR '99* (San Francisco, CA, USA, October 20– 21, 1999), pp. 145–148, 1999.

15. R. Entner. *Smartphones to Overtake Feature Phones in U.S. by 2011*. Nielsen Wire, 2010. [Online]. Available: http://blog.nielsen.com/nielsenwire/consumer/smartphones-to-overtake-feature-phones-in-u-s-by-2011/ [Accessed: August 18, 2010].

16. W3C Geo. Geolocation Working Group. 2010. [Online]. Available: http://www.w3.org/2008/geolocation/ [Accessed: August 19, 2010].

17. W3C. HTML Media Capture. 2010. [Online]. Available: http://www.w3.org/TR/capture-api/ [Accessed: August 19, 2010].

18. W. Lee and W. Woo. Exploiting Context-Awareness in Augmented Reality Applications. *Proceedings of ISUVR '08* (Gwangju, July 10–13, 2008), pp. 51–54, 2008.

19. T. Hofer, W. Schwinger, M. Pichler, G. Leonhartsberger, J. Altmann and W. Retschitzegger. Context-awareness on mobile devices - the hydrogen approach. *Proceedings of HICSS '03* (Hilton Waikoloa Village, Island of Hawaii, January 6–9, 2003), pp. 292–301, 2003.

20. E. Prasolova Forland, M. Divitini and A.E. Lindas. Supporting Social Awareness with 3D Collaborative Virtual Environments and Mobile Devices: VirasMobile. *Proceedings of ICONS '07* (Martinique, April 22–28, 2007), pp. 33–38, 2007.

21. Acrossair. Nearest Tube. 2009. [Online]. Available: http://www.acrossair.com/acrossair_app_augmented_reality_nearesttube_london_for_iPhone_3GS.htm [Accessed: August 18, 2010].

22. Theodolite. 2010. [Online]. Available: http://hunter.pairsite.com/theodolite/ [Accessed: August 18, 2010].

23. Layar. Augmented Reality Browser. 2010 [Online]. Available: http://layar.com/ [Accessed: August 18, 2010].

24. Metaio. Junaio. 2010. [Online]. Available: http://www.junaio.com/ [Accessed: August 18, 2010].

25. Wikitude. 2010. [Online]. Available: http://www.wikitude.org/ [Accessed: August 18, 2010].

26. WhereMark. 2010. [Online]. Available: http://www.wheremark.com/ [Accessed: August 18, 2010].

27. Sekai Camera. Beyond Reality. 2010. [Online]. Available: http://sekaicamera.com/ [Accessed: August 18, 2010].

28. OnLive. 2010. [Online]. Available: http://www.onlive.com/ [Accessed: August 18, 2010].

29. X. Luo. From Augmented Reality to Augmented Computing: A Look at Cloud-Mobile Convergence. *Proceedings of ISUVR '09* (Gwangju, South Korea, July 8–11, 2009), pp. 29–32, 2009.

30. V. Reynolds, M. Hausenblas, A. Polleres, M. Hauswirth and V. Hegde. Exploiting Linked Open Data for Mobile Augmented Reality. *Proceedings of W3C Workshop: Augmented Reality on the Web* (Barcelona, Spain, June 15–16, 2010).

A Platform for Mobile Collaborative Augmented Reality Game: A Case Study of "AR Fighter"

Jian Gu, Henry Been-Lirn Duh, and Shintaro Kitazawa

Abstract This chapter describes the implementation of networking features in a mobile augmented reality (AR) gaming environment. A prototype of a mobile AR collaborative game "AR Fighter" is developed. AR fighter employs the TCP/IP protocol to enable multiplayer functionality in a mobile AR environment. One phone acts as the server and the other as the client. The two phones communicate to each other via WiFi or Bluetooth connection. Players are able to manipulate their own virtual avatars to interact with virtual objects in the game. We also enabled Mac server features. Individual users can update their models by connecting to the Mac OS server which manages the tracking of high scores and the provisions of in-game incentives.

1 Introduction

The recent fusion of AR and mobile technologies will allow the creation of novel mobile AR applications. As the image processing algorithms and processing capabilities of mobile hardware significantly improve, mobile AR will become more common. In addition, the game industry itself has grown significantly in recent years. We explored the feasibility of building a mobile AR network game — "AR Fighter".

"AR Fighter" is a remote AR fighting table/card game (See Fig. 1) we developed for the Apple iPhone platform. The concept is derived from existing game titles

J. Gu (✉)
KEIO-NUS CUTE center, Interactive & Digital Media Institute,
21 Heng Mui Keng Terrace, #02-02-09, National University of Singapore
e-mail: idmgj@nus.edu.sg

L. Alem and W. Huang (eds.), *Recent Trends of Mobile Collaborative Augmented Reality Systems*, DOI 10.1007/978-1-4419-9845-3_7,
© Springer Science+Business Media, LLC 2011

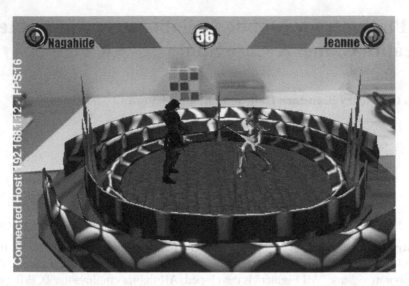

Fig. 1 AR Fighter

"Street Fighters"[1]. and "Eye of Judgment"[2]. The idea is combines mobile AR game with a common game. In the AR part, two users use different markers to virtually 'fight' each other. The mobile networking feature is added to this game to allow connected users to fight each other through the use of a common virtual object. A particle system is used to enhance the visual effects and a physics engine plays the essential role of detecting the collision of two virtual objects. User can download new model via our Mac server.

2 Related Works

In 2005, AR Tennis and The Invisible Train were developed as marker-based collaborative games on handheld devices. Henrysson ported ARToolKit to the Symbian platform and created a collaborative AR Tennis [1]. In AR Tennis, the smart phones are equipped with markers, which can be detected by the cameras of the other players phones. The tracking data is transmitted via a peer-to- peer Bluetooth connection, thus enabling the two players to play tennis virtually with their smart phones. The Invisible Train [4] is developed by Wagner and the aim is to steer a train over a wooden railroad track. The player can interact over the touch screen by changing the speed of the trains and the switches. The Invisible Train is a synchronized multiuser

---

[1] http://www.streetfighter.com/

[2] http://www.eyeofjudgment.com/

game in which PDAs are connected via WiFi. The tracking is realized by a marker-based computer vision approach. In 2009, Duy [3]discussed the potential uses and limitations of mobile AR collaborative games, and presented the Art of Defense (AoD), a cooperative mobile AR game. The AoD AR Board Game combines camera phones with physical game pieces to create a combined physical/virtual game on the tabletop. Int13[3] released a Mobile AR game called Kweekies, an AR pet training game that allows gamers to interact with their virtual pets by using the embedded cameras of their smart phones. Other existing applications are based either on the Windows or Symbian mobile operating platforms. Our AR Fighter is the first Apple iPhone OS AR collaborative game with full gaming features and Mac server support system.

3 System Structure of AR Fighter

The basic structure of our mobile AR game consists of two mobile phones that use a WiFi connection to link to each other for the AR game to be played. One phone acts as the server and the other acts as the client. We also established a server application in a Macintosh computer that acts as the server to manipulate and update the in-game models and to maintain high score tracking for the phone devices. When the iPhone application is started, it will check for the availability of updated models that a user may use and also list the high scores table. Fig 2 shows the structure of the game system.

4 Implementation

In almost all cases of implementations of AR functionality on mobile phones, developers usually would need to gain low-level camera data access for the required imaging technology to work properly. With the latest release of the iOS 4 update, Apple has allowed in-depth access to the iPhones camera API and therefore many new AR experiences are now possible. The latest version of ARToolkit[4] for iPhone helps us to achieve this feature. In addition, the Apple iPhone offers better real-time graphics performance than many other smart phones with the support of OpenGL ES 2, allowing for more complex applications to be designed and implemented. The collective supporting features formed the basis of our choice of the Apple iPhone as our development platform for our prototype work.

---

[3] http://www.int13.net/en/
[4] http://www.artoolworks.com/ARToolKitiPhone.html

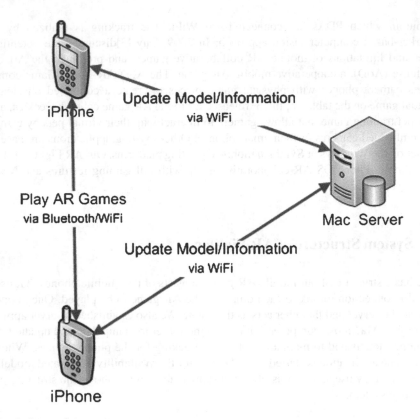

iPhone

Update Model/Information
via WiFi

Play AR Games
via Bluetooth/WiFi

Mac Server

Update Model/Information
via WiFi

iPhone

Fig. 2 System Structure

4.1 AR game functions

In this section, we describe the functions built for our AR game with a focus on network functionality.

AR library: We used ARToolkit for our development which was first released by ARToolworks[5] in 2009. This is the first fully-featured AR framework that supports native operations on various platforms. ARToolKits basic tracking works as follows:

- The camera captures a live video feed of the real world and sends it to the phone.
- Algorithms on the phone searches through each video frame for any square shapes.
- If a square is found, ARToolKit calculates the position of the camera relative to the black square.

---

[5] http://www.artoolworks.com/Home.html

- Once the position of the camera is known, a 3D model is drawn from that same position.
- This model is drawn on top of the video of the real world and so appears stuck on the square marker.

The final output is shown back in the phones screen, so the user sees virtual graphics overlaid on the real world.

3D animation loader: The loading and handling of resources is a complex part of a game. Numerous decisions have to be made about which file formats to support, how the files should be created and manipulated, the organization of data in these files and memory management issues pertaining to data that is loaded into the applications memory. We created a dynamic object loader for use with the OpenGL ES 3D graphics library. On the Apple iPhone, OpenGL ES runs on the embedded PowerVR hardware. Imagination Technologies[6] released a supplementary API to allow developers to take advantage of the PowerVR hardware. The model format for PowerVR is the POD format which supports node animations and bone animations. Animations are exported from 3DS MAX to POD files and the animation data for positioning, rotating and scaling is stored in each node that is animated within the scene file. The POD scene files loading performance in an AR environment is able to achieve a frame rate of 20-25 FPS while drawing an animated object consisting of 6200 polygons on the Apple iPhone 4. In this game, we preloaded eight geometry models for users to select from for the game (Fig 4).

Sound engine: There are generally two purposes for playing a sound in a game application, to serve as background music, and to serve as sound effects. Sound effect files tend to be short, ranging from 1 to 10 seconds long and do not take up much memory, partly because their sound quality can be lower than that of background music tracks. For our application, we included sound effects to correspond to gaming tasks in our AR Fighter game. However, as music files tend to be relatively longer in duration and larger in file sizes, they are usually compressed to reduce their file sizes. Unfortunately, that will also mean that such files will need to be uncompressed whenever playback is required and this step would require additional CPU usage. Since it is unlikely that multiple files will be played at any one time, this constraint is still manageable.

Physics engine: AR Fighter uses Bullet Physics for 3D physics simulation. Bullet Physics[7] is a library that supports 3D collision detection, soft body and rigid body dynamics and is available as open source.

Particle system: The 3D particle system is simply a group or groups of sprites that float about in virtual three dimensional space. However, for each particle instance, not only the geometry (if any), position, and orientation are tracked, but also its velocity, life span and animation speed as well. Real-time particles are added to enhance the visual effects.

---

[6] http://www.imgtec.com/
[7] http://bulletphysics.org/wordpress

4.2   Mobile AR network Functions

Networking support is an essential feature for a complete Mobile AR gaming experience. This is a vital component in our game. Wireless networking is necessary for communication between other users and systems while on the move, as defined in [2]. We designed two main structures for our network functions, a client-server architecture and a peer-to-peer architecture. In the client-server architecture, a Mac computer server updates the information for connected iPhones and it is assumed to have a known address so that the clients may conveniently connect to it with a pre-configured IP address. Players interactions and messages are sent from mobile clients to the server as they are generated. The server processes clients requests and facilitates automated processes such as the provisions of download services and the maintenance of the database of high scores. With a peer-to-peer model, each user would have to simulate the position of the other users virtual character, which will be explained as follows in fuller details.

Socket Class Hierarchy: Given a peer-to-peer architecture, each peer should be capable of initiating and hosting a game for some other peer to join, or join an existing game. Essentially, each peer would be able to function as a client (where it joins an existing game) or as a host (where it creates a game waits for another player). Initialisation of the network communication of a peer differs slightly depending on if the peer is a host or a client. For instance, a peer hosting a game has to listen for an incoming connection whereas a peer joining a game need not listen for anything. Therefore, each peer has two types of sockets available to it: the server socket and the client socket.

The server socket and the client socket shares a common interface known as the SuperSocket. This hierarchy is shown in Fig 3. The SuperSocket consists of methods that are common to both the sockets. It also consists of a series of methods whose implementation is provided by the subclasses. For instance, the send, recv, and isOffline methods need to be implemented by the subclass as it is dependent on how the socket is structured.

Mutual discovery: Without the presence of a central server with a known address, the discovering of a peers IP address becomes an issue. We solve this by allowing a mobile client to act as a game host and by revealing its IP address in the GUI. The second player can then connect to this game by entering the IP address of the host peer.

The disadvantage of our current approach is that both players need to have a communication channel outside of the game to communicate this IP address. It is not possible for two players to join each other just within the game without having to exchange IP addresses verbally or other such means. However, this is not a pressing problem as it is possible to create some kind of directory service where players advertise their IP address. The game can then retrieve the available games by querying this directory. Due to time constraints, we opted for a solution that is simpler to implement.

Performance: The network connection between the two players can be interrupted at any time. Once the connection is lost, we replace the lost player with AI so that

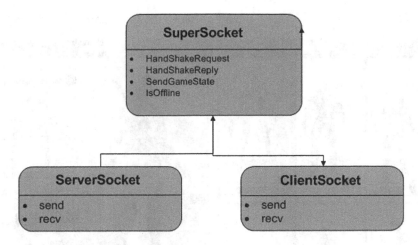

Fig. 3 Socket Class Hierarchy

game is able to continue. To achieve robustness, we maintain a flag to indicate the connection status. During each simulation step, we check the state of the socket and set the flag immediately when a disconnected is detected. We rely on the state of the TCP socket to know whether a disconnection has occurred and do not use any explicit heartbeat messages. The performance of Mobile AR Fighter game with the network support can reach 14-17 FPS.

4.3 User Interface design

In this paragraph, we will describe the details of the user interface design issues for AR Fighter. At the starting page, users need to determine if the mobile device will play as the host or as the client. They can then choose the character/avatar and arena as well as to see an overview of the opponents avatar (see Fig 4). Users can rotate the 3D avatar in the character viewport the choosing page by a finger swipe gesture across the screen of the mobile device.

In the AR interface, the accelerometer is used in our game to enhance game interactivity. The iPhones in-built accelerometer is a threeaxis device that it is capable of detecting either movement or the gravitational forces in three-dimensional space. User can tilt the phone to make the avatar move in the respective corresponding direction. We also used the multi-touch feature of the mobile phone in our game to allow players to control the virtual avatar. For example, a double tap makes the avatar move towards the corresponding point of the tap while a single tap is to initiate an attack move. Fig 5 shows a 3-hit which means that the user consecutively tapped the screen thrice. The vibration function is used whenever a virtual avatar is hit.

Fig. 4 Choosing the character

Fig. 5 Character is fighting

5　Conclusion

This chapter describes the development of a mobile AR collaborative game. We introduced the structure and features of the AR Fighter prototype. With a satisfactory system performance featuring a unique networked AR interface and interaction design, this brings an unprecedented mobile AR experience to users. Future work will mainly cover improvements in three areas:

1. Lighting sensitivity/condition is a common issue for AR applications. The relatively new Apple iPhone 4 features an embedded LED flash/bulb near its camera lens. We plan to use the camera API to control this strong flashlight to reduce the effects of external lighting conditions (resulting in possible over or under exposures in image captures) so as to preserve a consistent user experience by preventing disruptions in interaction flows.
2. Our game is currently designed only for 2 players. We aim to extend our networking feature to include additional player support within the network. Existing Wi-Fi network functionality will allow us to implement this feature easily. Our future tasks will include improving our mobile server part to support multiple connections and optimizing the frontend application to allow multiple players to play game at the same time with efficient 3D performance. Game session latency is another factor which needs to be considered when several players are added in the game. Currently, two players can play game by using Wi-Fi and Bluetooth with very little performance latency. This is because we are only sending relatively small data over the network. Considering data transfer latency, AR 3D performance in small screens and battery limitations of the phones, 4 simultaneous players play would be the maximum supported under this system framework. Finally, network traffic and security issues will also be studied at a more in-depth level.
3. Use of natural feature tracking feature to design interaction of avatars with finger gestures and real environmental objects. For example, users can control the avatar to dodge by hiding behind physical obstacles.
4. Wider range of supported platforms. The game will be playable across extended mobile platforms as a cross platform game.

Acknowledgements The initiative is supported by the Singapore National Research Foundation (NRF-2008-IDM-001-MOE-016) and the National University of Singapore (R-263-000-488-112). We specially thank Dr. Mark Billinghurst to help us acquiring ARToolKit commercial version for our work. Also thanks to Raymond Koon Chuan Koh to give us comments in the early draft and Yuanxun Gus help for implementing networking features.

References

1. Henrysson, A., Billinghurst, M., Ollila, M.: Face to face collaborative ar on mobile phones. In: ISMAR '05: Proceedings of the 4th IEEE/ACM International Symposium on Mixed and Augmented Reality, pp. 80–89. IEEE Computer Society, Washington, DC, USA (2005). DOI http://dx.doi.org/10.1109/ISMAR.2005.32

2. Huang, C., Harwood, A., Karunasekera, S.: Directions for peer-to-peer based mobile pervasive augmented reality gaming. In: ICPADS '07: Proceedings of the 13th International Conference on Parallel and Distributed Systems, pp. 1–8. IEEE Computer Society, Washington, DC, USA (2007). DOI http://dx.doi.org/10.1109/ICPADS.2007.4447813
3. Huynh, D.N.T., Raveendran, K., Xu, Y., Spreen, K., MacIntyre, B.: Art of defense: a collaborative handheld augmented reality board game. In: Sandbox '09: Proceedings of the 2009 ACM SIGGRAPH Symposium on Video Games, pp. 135–142. ACM, New York, NY, USA (2009). DOI http://doi.acm.org/10.1145/1581073.1581095
4. Wagner, D., Pintaric, T., Schmalstieg, D.: The invisible train: a collaborative handheld augmented reality demonstrator. In: SIGGRAPH '04: ACM SIGGRAPH 2004 Emerging technologies, p. 12. ACM, New York, NY, USA (2004). DOI http://doi.acm.org/10.1145/1186155.1186168

Effect of Collaboration and Competition in an Augmented Reality Mobile Game

Leila Alem, David Furio, Carmen Juan, and Peta Ashworth

Abstract Greenet is an augmented reality mobile game developed for children for learning about how to recycle. In this paper we present a study involving 28 primary school students, to explore the extent to which collaboration and competition affect perceived learning and potentially lead to a change of attitude and behaviour. In this study, students in sessions of 4 were asked to play recycling games in 3 conditions: by themselves, in a team, and in a team while in competition with another team. Our results show that collaboration and competition promotes a positive change of attitude and behaviour. This study suggests that competitive/collaborative mobile phone based games provide a promising platform for persuasion.

1 Introduction

In recent years, mobile phones have been used to reach and motivate people to change their attitudes and behaviours. Unlike desktop computers, mobile phones are ubiquitous and portable. They offer immediacy and convenience, they go wherever one goes and hence are seen as a good candidate for persuasive technology [3]. A number of applications have been developed on mobile phones in order to reach and motivate people to change their attitudes and behaviours particularly around health and the environment. For example, the SexINFO service is a sexual health service for young people that is accessible via mobile phones [4] while BeWell Mobile is an asthma management application for children and teens with severe asthma [1]. Similarly, Ubifit Garden and Environmental Awareness are two applications developed by Intel research to persuade people to change their attitude and

L. Alem (✉)
CSIRO ICT Centre, PO Box 76, Epping NSW 1710, Australia
e-mail: Leila.Alem@csiro.au

L. Alem and W. Huang (eds.), *Recent Trends of Mobile Collaborative Augmented Reality Systems*, DOI 10.1007/978-1-4419-9845-3_8, © Springer Science+Business Media, LLC 2011

behaviours in their daily life [2]. Ubifit Garden encourages people to increase their daily physical activity, by displaying on the screen of the phone a garden that booms as people perform their activities throughout the week. A butterfly appears when meeting their weekly goal. Alternatively, Environmental Awareness is a mobile phone application that uses sensors and GPS to measure property of air quality. Real time air quality reports are generated and distributed to other mobile phone users.

While all these mobile phone applications have one objective in common, that is to motivate change in people's attitude and behaviours, very little is reported in actually explicitly assessing/measuring the potential persuasive power of these mobile applications.

In this paper we present our initial findings of a study using an augmented reality (AR) mobile game designed to promote learning and a positive change of attitude about recycling. In our study we are interested in investigating the individual effect of collaboration and competition on persuasion and learning. While collaboration and competition are two important intrinsic motivation factors, a review of the literature on mobile persuasion showed that no study has attempted to investigate and quantify the effect of collaboration and competition on persuasion. In our study we use Greenet, a mobile phone game focused on recycling using a Nokia N95.

In the next section we present our augmented reality mobile phone games in detail, followed by the results of a trial conducted in May 2009. These results are discussed and a number of conclusions and areas for future research are identified.

2 Greenet Game

Greenet is an augmented reality game for mobile phones dedicated to recycling. The game uses one mobile phone, one object marker and 4 recycling bin markers. When the camera of the mobile phone detects the object marker, a recycling object is displayed on the screen. When the phone detects a bin marker, a bin is displayed on the phone's screen. The game consists of picking up an object by selecting the "OK" button of the phone when the object is displayed on the screen, and dropping the object in the right recycling bin. The player has to look for the bin markers in order to identify which bin the object needs to be placed. When the correct bin is shown on the screen, the player presses the "OK" button to place the object into the bin. The objects to recycle are randomly selected; the bins markers allow the display of four bins consisting of the correct bin plus three other bins that are randomly selected.

Players earn points for putting the objects in the right recycling bins and lose points otherwise. Players are also presented with multiple choice questions. Players earn points for providing the right answer to questions and lose points otherwise. Visual feedback is provided for every action of recycling and after answering every recycling question. Clapping hands are displayed for successful actions and a red cross otherwise. The game consists of two levels, level 1 and level 2. At level 1, players recycle few objects using only 2 recycling bin markers and the recycling questions are selected in easy to moderate level of difficulty. At level 2, players recycle more objects using 4 recycling bin markers and the questions are more difficult.

Fig. 1 Nokia N95

Developing the Greenet game requires a mobile phone with an integrated camera and hardware graphics acceleration with enough power to play the game at a reasonable speed. The mobile phone that fulfils the requisites at that moment was the Nokia N95 8Gb using Symbian 9.2 OS. To develop our game application on this phone the researchers used Microsoft Visual Studio 2005 with the plugin Carbide. vs and S60 3rd Edition Feature Pack 1. For AR capabilities ported ARToolKit 2.65 onto the phone was used.

In this paper we explore the extent to which collaboration and competition promotes a positive change of attitude and behaviour.

3 The Study

Participants played the Greenet game in the following three conditions:

Basic Greenet: in this condition 4 players are playing solo e.g. without any interaction with other players. Each player while playing can see their performance and at the end of their game can see their performance in relation to other players' performance on their Nokia screen.

Collaborative Greenet: in this condition the 4 players are grouped in pairs. Each player plays with their partner, there is no competition in this condition, just collaboration. Each player while playing can see their team performance and at end of game can see their team performance in relation other teams' performance on their Nokia screen.

Competitive Greenet: in this condition the 4 players are grouped into two competing pairs. There is both competition and collaboration in this condition. Each player while they are playing can see their team performance and their opponent team's performance on their Nokia screen. At the end of the game they can see their team performance in relation of other team's performance on their Nokia screen.

In this study we used a mixed method approach for data collection, including questionnaires, interviews, task performance and written recounts of participants' experiences of being involved in the study.

The underlying assumption in this research is that all intrinsic persuasive factors working together will result in enhanced persuasion outcomes than the factors taken separately.

Competition > collaboration > basic for persuasion and perceived learning scores

3.1 Participants and Procedure

Twenty eight (28) children from a local primary school in Sydney took part in this study. In total there were sixteen (16) girls and twelve (12) boys. The average age of children was ten (10) years old (m= 10.18).

Of the sample, over 53% of the children used mobile phones once a week or more and 76% of the children reported playing computer games once a week and more. Almost 86% of the children felt quite knowledgeable about what can be recycled and 75% about how to recycle. Of the total sample, 86% of the children stated they already recycle at home and 93% of the children agreed that people should recycle in order to reduce their environmental footprint.

Participants were collected from their classroom and taken to the school library to participate in the game. Firstly the participants were familiarized with the game and the mobile phone once they had completed a consent form. Participants were then asked to fill in an Entry questionnaire and then played the game under each of the 3 condition – basic, collaborative and competitive - the conditions were randomized. Once they had played the game, the participants were asked to fill in a Post questionnaire after each condition. After completing all 3 conditions, participants were asked to fill in an Exit questionnaire, then they were ask to describe their collaborative experience. Upon completion of the session, the participants were returned to their individual classrooms.

Each session involved four participants and lasted up to 40 minutes. Groups of four participants were formed in advance in order to minimize disruption to the class. Groups were generated using first alphabetical order and were then reviewed in order to balance gender where possible. In this study 7 sessions were run, requiring a total of 28 participants.

4 Results

4.1 Perceived Learning Value

In answer to the question, I think playing this game has helped me learn about residues that can be recycled (1=None, 4=Some, 7=A great deal). The majority of children found the three conditions of value from a learning perspective, with a higher

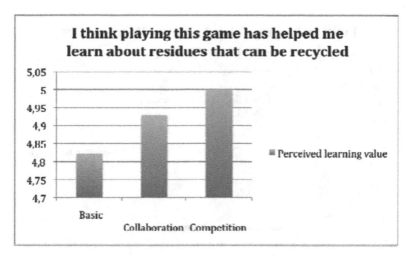

Fig. 2 Perceived learning value

Fig. 3 Attitudes

mean score for Competition (m=5, std=1.68) over collaboration (m=4.93, std =1.96) and Basic score (m=4.82, std=2.13).

4.2 Attitudes

Participants were also asked to respond to the question "People should be recycling more in order to reduce their environmental footprint (1= Strongly disagree 4= Unsure and 7= Strongly agree)." In general the children demonstrated a strong

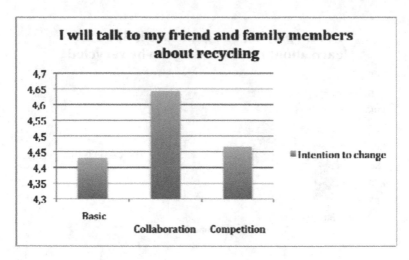

Fig. 4 Intention to change

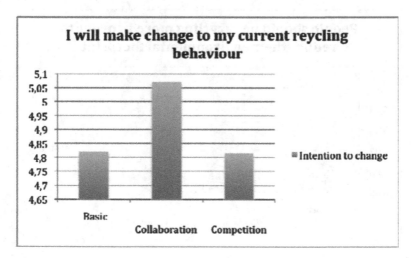

Fig. 5 Intention to change

attitude towards recycling, with a higher mean score for Competition (m=6.11, std=1.45) and collaboration (m=6.11 std =1.29) over Basic score (m=6.07, std=1.64).

4.3   Intention to Change Behaviours

Participants were asked a number of questions about their intention to change behaviors. In response to the question "I will talk to my friends and family members

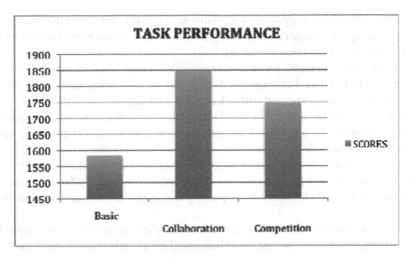

Fig. 6 Scores

about recycling (1= None, 4= Some, 7= A great deal)" Participants showed positive support for the idea of talking to friends and family. The highest response was for the Collaboration (m=6.64, std=1.91), condition, followed by Competition (m=4.46, std=1.93),and Basic (m=4.43, std =1.99).

Finally, in response to the question: I will make changes to my current recycling behaviour (1= None, 4= Some, 7= A great deal)." The Collaboration condition appeared to gain the highest positive response (m=5.07, std=1.70) compared to both competition and basic conditions.

4.4 Performance scores

A higher task performance score for Collaboration (m=1847.64, std=387.40) over competition (m=1746.5, std=346.16) over basic score (m=1584.57, std=682.92) is reported. A significant difference between the task performance scores of the collaboration and basic condition is reported (r= 0.0263, p<0.05).

5 Discussion

In accordance with our hypothesis, the competition condition is perceived as the condition in which more learning about recycling is taking place. On a more objective level, students were performing significantly better (doing the right recycling actions and providing the right answer to recycling questions) in the collaboration condition over the basic condition.

We hypothesised competition> collaboration >basic for intention to change scores, and report collaboration > competition > basic for all means scores of the two dimensions of intention to change.

Extracts from students' recount records, when asked ways in which the game may have affected their behaviour "I think I will start recycling more" "yes I have learn more about recycling and will talk about it with anyone who will listen" "It did not make me want to recycle but it did teach me about what to recycle" "we learnt to cooperate well" "yes it has but we already recycle at home" "it get me to see if anything can be recycle after use".

Some students even reported intentions to change" I have talked to my parents and how I will take out the recycling when needed to" " I want my family to recycle more and I want to do it" "yes I think playing this game has change my mind about recycling".

These comments seem to suggest that the AR game presented in this paper has the potential to influence people into changing their attitude towards recycling.

6 Conclusion

Our results suggest that collaboration and competition promotes a positive change of attitude and behaviour. Greenet, the game presented in this paper, is a game in which players learn about recycling by practicing the act of recycling using a mobile phone. This study suggests that competitive/collaborative mobile phone based games provide a promising platform for persuasion.

In future work we plan to further examine participants' knowledge and attitudes towards recycling and the environment in order to identify if this impacted on their ability to perform the recycling behaviours required in the AR game.

References

1. Boland, P. The emerging role of cell phone technology in ambulatory Care. Journal of Ambulatory Care management, 30, 2 (2007), 126–133.
2. Consolvo, S., Paulos, E., and Smith, I. Mobile Persuasion for Everyday behavior change. In Mobile Persuasion: 20 perscpectives on the future of behavior change. Eds Fogg BJ and Eckles D. (2007), 77–84.
3. Fogg, B.J. Persuasive Technology: using computers to change what we think and do. Morgan Kaufmann Publishers. An inprint of Elsevier. ISBN-10:1-55860-643-2 (2003).
4. Levine, D. Using Technology to promote Youth Sexual Health. In Mobile persuasion : 20 perspectives on the future of behavior change. Eds Fogg BJ & Eckles D (2007); 15–20.
5. Rohs, M. Marker-Based Embodied Interaction for Handheld Augmented Reality Games, Journal of Virtual Reality and Broadcasting, 4, (2007), no. 5.

A Collaborative Augmented Reality Networked Platform for Edutainment

Yuan Xun Gu, Nai Li, Leanne Chang, and Henry Been-Lirn Duh

Abstract This chapter presents a mobile software prototype for the educational purpose by using the Augmented Reality (AR) and data communication technology. We developed a prototype of mobile game called "AR-Sumo" that aims to offer a shared virtual space for multiple mobile users to interact simultaneously. "AR-Sumo" invloves visualization of augmented physical phenomena on a fiducial marker and enables learners to view the physical effects of varying gravities and frictions in a 3D virtualspace. The software implementation is packaged as network service where it is broadcasted from an access point connecting to a designated server. With the established network connection, mobile phones users can receive services semi-ubiquitously within the range of the broadcast. This architecture resolves the issue of heterogeneity of computational capacity among different types of mobile phones. As an on-going project, future studies will focus on the usability and designs with an attempt to enhance the efficacy of the application.

1 Introduction

AR is a technology that combines the virtual scene with reality. It is a multidisciplinary area that has been developed for decades, focusing on vision tracking, interaction technique and display technology [19]. Early works on collaborative AR [6] were mostly implemented on desktop computers so it restricted the mobility of AR applications. However, this restriction has been removed in recent years when AR applications can be implemented by portable devices such as certain smart phones (e.g. iPhone), which are equipped with the capacity to process the required computation.

Y.X. Gu (✉)
Department of Electrical and Computer Engineering, 21 Heng Mui Keng Terrace,
#02-02-09, National University of Singapore
e-mail: yuanxun@nus.edu.sg

L. Alem and W. Huang (eds.), *Recent Trends of Mobile Collaborative Augmented Reality Systems*, DOI 10.1007/978-1-4419-9845-3_9,
© Springer Science+Business Media, LLC 2011

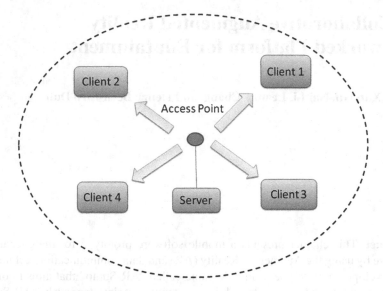

Fig. 1 Concept of Semi-ubiquitous AR Service

It is a known concept that 3D visualization helps to develop human spatial ability [7]. AR technology is a good medium for establishing a mobile environment. The concept of collaborative AR learning through shared augmented space has been investigated in [2, 8, 17, 19]. Their interface designs allow users to engage in collaboration more intuitively (i.e., in real-time space) but the setups require intensive supports of hardware. Also, the ubiquitous computing system provides extensive mobility or maneuverability. AR tennis [4] is a good example of a portable AR collaboration for entertainment. The objective of this project is to incorporate collaborative AR into mobile learning system. A collaborative game that is designed to be educational would transform the way people learn new knowledge. Our targeted mobile users are students of primary school and we would like to examine how this alternative learning approach would impact the way in which students learn about the physical world.

In this chapter, we describe a proof-of-concept software prototype of "AR-Sumo". The system demonstrates the feasibility of developing a server-based collaborative mobile AR application. By design, we are interested in offering AR collaboration as a network service (AR service) targeting to serve mobile clients. Service is provided by a dedicated server that is attached to a Wi-Fi access point (Fig. 1), which broadcasts service within a small space such as classroom, office, and multi-person workspace, etc. It allows mobile phones that are within the network proximity to receive AR service through their Wi-Fi connection with the server. The concept is motivated by scenarios in which AR services can be offered ubiquitously from neighbourhood stores, restaurants and classrooms, functioning as a novel way of delivering advertisement and education. Such design not only supports AR processing in low-powered mobile phones but also facilitates easiness of content upgrading at the server side to support more advanced features (e.g. implementation of more intensive application simulation) without the concern of the heterogeneity issue of computational capacity among different mobile phones. For educational purposes, classroom environment is one of the ideal places to broadcast the AR service because it allows students to achieve interactive learning collaboratively.

Section 2 of this chapter covers our reviews on collaborative mobile learning and the theoretical guidelines to the design of "AR-Sumo". Section 3 gives a detailed discussion of our system implementation in technical aspects. We conclude this chapter with some discussion on our future works in the final section.

2   Mobile AR & Learning

2.1   Past Collaborative AR Works

Early works on collaborative AR focused on head-mounted display (HMD), desktop and handheld-based environment. Construct3D [6] is designed as a 3D geometric construction tool that can be used for a wide range of educational purposes. Students wearing HMD can engage in face-to-face interactions in real-time 3D virtual space. Similarly, AR Tetris [18] allows users to collaborate remotely with fiducial markers in a master/trainee scenario. These collaborative systems are designed to be applied in a range of educational contexts. However, the investment-intensive hardware requirement makes them impractical to be widely deployed outside the research laboratory. ARQuake[15] is a mobile AR indoor/outdoor application that uses both GPS information and vision based technique. It is enabled by a backpack configuration in which its cost and performance (30 frames per second) are balanced. In contrast, AR tennis [4] is designed for mobility as in the expensive AR computation and game simulation are both processed internally in mobile phones and no additional external hardware is required. Although fully functional, its pitfalls are its' low resolution in augmented video frame and slow frame transition rate (i.e., 3 to 4 frames per second). To overcome these pitfalls, our "AR-Sumo" prototype is designed as a semi-ubiquitous architecture because of the additional server. The superb data transmission speed from a stable and strong Wi-Fi connection gives the average performance of 10 frames per second. This archived performance outperforms AR tennis significantly. Our "AR-Sumo" has avoided the pitfalls of both AR Tetris and AR Tennis face, and it has achieved a relatively good application performance.

2.2   Overview of Mobile Learning Design

The relevant research on collaborative mobile learning environment design is first reviewed to guide the software prototype development and analysis. Mobile learning enables individuals to access greater educational information via mobile devices without the constraints of time and place. Such practicality contributes to richer learning experiences for individuals. However, the mobile technology may be an impediment to the learning process if the design cannot fit the context. To address the effectiveness issue on the application aspect of mobile learning, researchers have provided insights into the requirements for applications of mobile learning environments by considering multiple factors. Parsons, Ryu and Cranshaw [11] developed a framework for designing mobile learning environments by integrating three factors, namely, learning context, learning objectives and learning experiences.

The collaborative task is an integral part in designing mobile technologies. In the educational context, collaborative learning, referring to two or more individuals learning some new information or knowledge together, is identified as an effective approach for enhancing learning effectiveness [14]. As the advancement of mobile technologies keeps revolutionizing the collaborative learning environment, design issues related with mobile collaborative learning systems have been received considerable attention in an effort to promote learning quality in recent years.

Three principle aspects should be taken into account for evaluating the efficacy of the design. The technological aspect for the design revolves around usability issues such as user interface, hardware, and software systems. Multiple dimensions are incorporated into the usability assessment which includes controllability, learnability, satisfaction, feedback, menu/interface, etc. [5]. Specifically in the education domain, pedagogical and ffectogycal influences should also be considered [7]. The pedagogical aspect pertains to the ffecttiveness of delivering knowledge to users. And the psychological aspect refers to the engagement and preference of the users. Mobile AR is an emerging technology that enables users to see the real world augmented with virtual objects by utilizing mobile devices. To enhance the usability and effectiveness of mobile AR applications in education setting, researchers have directed at their investigation on the physical configuration and virtual content, and proposed strategies for achieving greater efficacy in the AR environment [7, 13].

An interface with good usability should naturally support human-computer interaction, which allows users to manipulate the virtual objects as intuitively as possible. Additionally, the design and the performance of the objects on the interface that execute the user's manipulations should be user friendly [1]. In terms of designing the content, the main focus is how well it would facilitate learning. The cinematic and the game metaphors are helpful for strengthening learners' engagement in the mobile learning environment [11].

In summary, technical design, pedagogical and psychological aspects are crucial in making a mobile AR-supported educational application provide successful and pleasant experiences for the users. The technical usability and educational effectiveness are the two most important domains to warrant for a careful attention when designing a mobile AR environment with the aim of fostering collaborative learning. Also, it is good to integrate some entertaining elements to keep users' attention and engagement. This would strengthen the effectiveness of the delivery of educational content.

Fig. 2 Game View (AR Sumo)

2.3 Game Design

Mobile AR technology entails potentials in providing an optimal collaborative learning environment. The combination of spatial and tangible interface of the mobile AR application creates opportunities for individuals to directly manipulate virtual objects in the physical environment whilst engage in a face-to-face communication [2]. AR technology can also be applied to construct simulations with a high degree of realism. It is especially useful to facilitate the understanding of science phenomena that are difficult to observe in physical life unless utilizing complex equipments, and in turn fosters knowledge acquisition [12].

Hence, game-based collaborative learning supported by mobile AR technology enables individuals to learn things more coherently as they could visualize the relevant elements, as well as more socially as they have to interact with their peers. They can actively seek and construct knowledge, while the immersion in the game can stimulate a sense of engagement, hence promoting learning experience [3, 13].

Recognized the capacity of mobile AR technology in edutainment, our collaborative "AR-Sumo" game (Fig. 2) is designed by the abovementioned principles. As an educational game, "AR-Sumo" allows learning while playing. Motivated by a 2D educational sumo game [9], we have added an additional dimension of visualization and control feature in our "AR-Sumo" game, and also managed to register the entire virtual game world into the reality. "AR-Sumo" allows the registrations of two individuals to play the game. Each player is able to control a 3D rectangular virtual block (Fig. 2). The lesser the collisions with boundaries of the virtual world indicates the better the player performs. The player can choose to use a virtual block controlled by him/her to hit his/her counterpart. The intention of hitting is to sabotage the counterpart to collide into the boundary of the virtual space. If the total number of collisions recorded exceeds a pre-determined value, the player is declared the loser. On the server side, a control console is used to manipulate several virtual world parameters. For instance, a facilitator (e.g. a teacher) can choose to manipulate (tune in, turn on or off) the degree of gravity and friction of the virtual world, which would influence players' experience. Players are made aware that the varying degree of gravitation and friction through their own visualization in the virtual world.

The AR interface should be intuitive for learners to use. In our application, two rectangular blocks with different colors are assigned to two users so that each user can control a block. In order to offer intuitive control interface, a total of six arrow-shaped buttons are located at the bottom right side of the screen (Fig. 2) as the indication of applying force to the assigned block in one of the six primitive directions. (i.e. +/- x, +/- y, +/- z). In addition, the player's viewpoint toward the virtual 3D space can be easily changed by moving the phone around.

3 Implementation

3.1 System Architecture

Typical AR applications are sequential combination of video frame acquisition, fiducial marker detection, application/game simulation and graphic rendering. Depending on the computational capabilities of the respective mobile devices, it would be relatively expensive to accommodate the entire sequential combination of tasks. User experience will be reduced accordingly by application performance. The solution is to offload certain task(s) to a server with powerful processing capability.

Four different designs of user-server architectures have been proposed [16]. The first design was a handheld self-contained structure in which the entire AR computation could be processed in a mobile device. The experiments on Google Nexus one phone showed that this approach yielded poor-quality augmented video (i.e., 5.1 frames per second on average). The second design allowed the task of fiducial marker detection to be offloaded to the server while keeping application simulation and graphic rendering in mobile device. This design is logical. However, having application/game simulation on the same mobile device would cause a huge inconsistency in the application/game state among different mobile devices. Similarly in the third design, the server took over the task of application simulation and became a central processing unit for the AR service.

The last design involved a two-way video transmission and it required a high bandwidth network. While it appeared to be the easiest solution to offload all expensive tasks to server, the disadvantages are palpable. The stability of a Wi-Fi environment could hardly provide a good quality of service (QoS) due to the extremely heavy demand on the network bandwidth for a continuous two-way real-time multimedia content delivery.

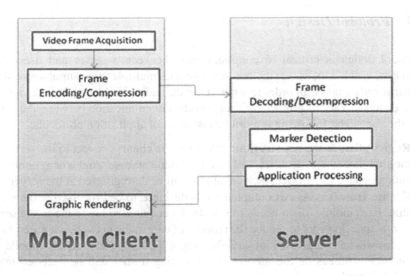

Fig. 3 System Architecture

While [16] chose to use a combination of the first and second design in implementing their handheld AR system because of the requirement of permanent presence of server. Our software prototype favoured the third design in order to give the server the required capabilities to handle the AR computation and application simulation (e.g. physics simulation). That is, mobile clients would acquire and compress their video frames, and send them to a server for further processing. The dedicated server which receives these video frames would decode them for fiducial marker detection, simulate visual environment states and return rendering commands to each mobile client that has initiated the AR service. Upon receiving commands from the server, mobile users render augmented frames accordingly. The commands are sent continuously from the server, which consists of the model-view matrices of the virtual world's boundary and each virtual object.

3.2 Physics Engine

A physics engine has been implemented on the server to offer the physical capability to 3D objects in the virtual world. In our design, the rate of physics simulation is coupled to the rate of frame arrival at server in order to make the game simulation speed adaptive to the variable frame arrival rate at the server side. The engine enables rigid body collision detection within the virtual world and visualizes the effect of applying force to the virtual objects in a user-controlled 3D environment. In addition, friction and gravitation are the control parameters offered by the engine, and it could be easily tuned in at the server's control console.

3.3   Protocol Design

Protocol design is critical to communication between a server and users. The prototype enabled the server to communicate with multiple users simultaneously to perform certain tasks. In order to support the desired task in our system design, we have designed a set of protocols to support the communication between the server and the users. The following is a brief illustration of application protocols:

- **Registration & De-registration:** When mobile clients connect to the server for the first time, their identities will be added into database. And when mobile clients terminate the service, their identities will be de-registered at the server.
- **Frame Transfer:** As AR computation is designed to be processed at the server side. Each mobile client needs to transfer its acquired video frame to the server in real time. In order to reduce the amount of data in the network, captured video frame will be compressed into a light weight YUV420 format at the mobile client side; whereas on the server side, incoming frames will be decoded into a RGB format.
- **AR Processing:** "AR-Sumo" employs NyARToolkit [10] to detect fiducial marker and it returns 4×4 transformation matrix (model view matrix) facilitating OpenGL to draw 3D virtual world onto the fiducial marker. Transformation matrices of the rest of the virtual objects are then the product of multiplication between model view matrix and the rigid body transformation matrices (e.g. translation and rotation) that derived from the physical simulation.
- **User Interaction:** Differ from the AR processing as a real-time periodical event. Network messages resulted from player interactions are only exchangeable when a player event is initiated. Messages will be delivered to the server and be simulated from a subsequent simulation cycle.

4   Conclusion & Future Work

Mobile AR plays a significant role in facilitating collaborative learning. This chapter presents "AR-Sumo", which is a mobile collaborative augmented reality network service for educational and entertainment purposes. An important application of this system is to support existing collaborative learning in school setting. The selected educational elements (e.g. applying force to object from different directions, the effect of different degree of gravitation and friction) in physical world are conveyed to the players throughout their manipulation of virtual objects during the game.

The present work enriches the research on mobile collaborative AR in edutainment domain. First, it creates more opportunities for students to engage in collaborative learning through taking advantages of potentials of AR technology in education. Second, the application is implemented by mobile phones, which allows users to interact with each in natural styles without the constraint of fixed physical setups. Third, the novel semi-ubiquitous AR service architecture is designed to treat AR as network services. This concept can be applied not only in classroom setting but also in many other similar scenarios.

As an on-going project, the development works for the subsequent phase are still in the progress. The current version of software prototype and game design support the collaboration of two players within the broadcast range at an average performance of 10 frames per second on mobile phone. The client software has been deployed to Google Nexus One phone and the server program is situated at a workstation connecting to a router. We would like to support more users to collaborate at the virtual space simultaneously in the near future.

Other future works consists of two main directions: It is necessary to design and conduct a study on user experience of the system. We will evaluate the effectiveness of "AR-Sumo" in terms of knowledge delivery, engagement of users and the usability of the system with an attempt of exploring better evaluation methods of mobile collaborative AR systems. Second direction is to improve its robustness against network congestion. We are now investigating a local dead reckoning algorithm that make use of the information from mobile phone's inertia sensor and acquired video frame to predict marker position.

Acknowledgements This initiative is supported by the Singapore National Research Foundation (NRF-2008-IDM-001-MOE-016) and the National University of Singapore (R-263-000-488-112).

References

1. Apted T, Kay J, Quigley A (2005) A study of elder users in a face-to-face collaborative multi-touch digital photograph sharing scenario. Technical Report Number 567, The University of Sydney
2. Billinghurst M, Poupyrev I, Kato H, May R (2000) Mixing realities in shared space: An augmented reality interface for collaborative computing. Paper presented at the IEEE International Conference on Multimedia and Expo, New York, USA, 30 July-2 August
3. Clarke J, Dede C, Dieterle E (2008) Emerging technologies for collaborative, mediated, immersive learning. In: Voogt J, Knezek G (eds.) International Handbook of Information Technology in Education. Springer, New York
4. Henrysson A, Billinghurst M, Ollila M (2005) Face to face collaborative AR on mobile phones. Paper presented at the IEEE/ACM International Symposium on Mixed and Augmented Reality, Vienna, Austria, 5-8 October
5. Kaufmann H, Dünser A (2007) Summary of usability evaluations of an educational augmented reality application. Paper presented at the HCI International Conference, Beijing, China, 22–27 July
6. Kaufmann H, Schmalstieg D, Wagner M (2000) Construct3D: A virtual reality application for mathematics and geometry education. Educ Inf Technol 5: 263-276. doi: 10.1007/s10639-010-9141-9
7. Kaufmann H (2003) Collaborative augmented reality in education. Paper presented at the Imagina Conference, Monte Carlo, Monaco, 3 February
8. MacWilliams A, Sandor C, Wagner M, Bauer M, Klinker G, Bruegge B (2003) Heading Sheep: Live System Development for Distributed Augmented Reality. Paper presented at the IEEE/ACM Symposium on Mixed and Augmented Reality, Tokyo, Japan, 7–10 October

9. MyPhysicsLab-Sumo Wrestling Game.http://www.myphysicslab.com/collisionGame.html. Accessed 5 June 2010
10. NyARToolkit for Java.en http://nyatla.jp/nyartoolkit/wiki/index.php?NyARToolkit%20for%20 Java.en. Accessed 4 June 2010
11. Parsons D, Ryu H, Cranshaw M (2007) A design requirements framework for mobile learning environments. JCP 2: 1-8. doi: 10.4304/jcp.5.9.1448–1455
12. Reamon D T, Sheppard SD (1997). The role of simulation software in an ideal learning environment. Paper presented at the ASME design engineering technical conferences, Sacramento, CA, USA, 14-17 September
13. Schrier K (2006) Using augmented reality games to teach 21st century skills. Paper presented at the ACM SIGGRAPH Conference, Boston, Massachusetts, 30 July 30–3 August
14. Smith BL, MacGregor JT (1992) What is collaborative learning? In: Goodsell AS, Maher MR, Tinto V (eds.), Collaborative Learning: A Sourcebook for Higher Education. National Center on Postsecondary Teaching, Learning, & Assessment, Syracuse University
15. Thomas B, Close B, Donoghue J, Squires J, De Bondi P, Morris M, Piekarski W (2000) ARQuake: An Outdoor/Indoor Augmented Reality First Person Application. Paper presented at the International Symposium on Wearable Computers, Atlanta, GA, USA, 16-17 October
16. Wagner D, Schmalstieg D (2003) First steps towards handheld augmented reality. Paper presented at the IEEE International Symposium on Wearable Computers, New York, USA, 21–23 October
17. Wagner D, Pintaric T, Ledermann F, Schmalstieg D (2005) Towards massively multi-user augmented reality on handheld devices. Paper presented at the International Conference on Pervasive Computing, Munich, Germany, 8–13 May
18. Wichert R (2002) A Mobile Augmented Reality Environment for Collaborative Learning and Training. Paper presented at the World Conference on E-Learning in Corporate, Government, Healthcare, and Higher Education, Montreal, Canada, 15–19 October
19. Zhou F, Duh HBL, Billinghurst M (2008) Trends in augmented reality tracking, interaction and display: A review of ten years of ISMAR. Paper presented at the IEEE/ACM International Symposium on Mixed and Augmented Reality, Cambridge, UK, 15–18 September

Prototyping a Mobile AR Based Multi-user Guide System for Yuanmingyuan Garden

Yongtian Wang, Jian Yang, Liangliang Zhai, Zhipeng Zhong, Yue Liu, and Xia Jia

Abstract Based on the recent developments in the field of augmented reality(AR) and mobile phone platform, a multi-user guide system for Yuanmingyuan Garden is designed. The proposed system integrates real environment and virtual scene through the ways of entertainment and gaming in the mobile phone, so that the brilliant royal garden in the ancient time can be rebuilt onsite. By the interactive ways of scene reconstruction and scenario setting, its users can not only be guided to appreciate the current views of the Yuanmingyuan Garden, but also experience the glory and the vicissitudes of the ancient Chinese garden cultures. The proposed system brings novel ways of entertainment to the sightseeing trip of the visitors, and the visitors will be benefited from the learning of the cultural sites and the cultural histories. In this paper, the prototype of the mobile AR application of Yuanmingyuan garden has been presented. The proposed system is still at its early stage and requires further development. However, the novel concepts proposed in this system will bring people with inspirations for the integration of advanced AR technology to the tourism industry.

1 Introduction

Yuanmingyuan Garden, which is called the garden of all gardens, is located in the northwest of Beijing. It was built from Kangxi period of Qing Dynasty in the early 18th century and covered a total area 1 of 350 hectares. Yuanmingyuan Garden was an exquisite royal palace created by Qing emperors through a century and was the epitome of Chinese ancient palaces and gardens, known as the "Eastern Versailles".

Y. Wang (✉)
School of Optics and Electronics, Beijing Institute of Technology, Beijing 100081, China
e-mail: wyt@bit.edu.cn

L. Alem and W. Huang (eds.), *Recent Trends of Mobile Collaborative Augmented Reality Systems*, DOI 10.1007/978-1-4419-9845-3_10, © Springer Science+Business Media, LLC 2011

Fig. 1 The virtual reconstruction of Yuanmingyuan Garden

Unfortunately, Yuanmingyuan Garden was looted and burnt down by the Anglo-French allied forces in the October of 1860, followed by innumerable devastation. The famous garden went into ruins eventually. Nowadays, the ruins of Yuanmingyuan Garden are the precious historical and cultural heritage of Chinese nation.

Because of its historical position, the reconstruction of Yuanmingyuan Garden is controversial all the time. But the AR-Based [1] virtual reconstruction realized by modern technology is more meaningful. Otis lab in Beijing Institute of Technology [2, 3] has developed the augmented reality based system of Yuanmingyuan Garden running on PC. In this paper, we present our early work on "AR virtual reconstruction of Yuanmingyuan " (as shown in Fig. 1).The display of the ruins is integrated with the digital virtual palace model not only makes visitors feel the site status quo, but also brings them the history with past glories. The visitors of Yuanmingyuan Garden can appreciate its value from the contrast of the reality and history.

In this paper, based on the virtual digital reconstruction combined with the mobile phone platforms, an entertainment system that involves multi-user and integrates the reality and virtual scences has been designed. By scene construction and scenario setting, the proposed system can allow its visitors to participate in the process of the virtual reconstruction of Yuanmingyuan Garden. The visitors can not only be guided to appreciate the scenery of the Yuanmingyuan Garden, but also can experience the glory and the vicissitudes of the Yuanmingyuan Garden from entertaining.

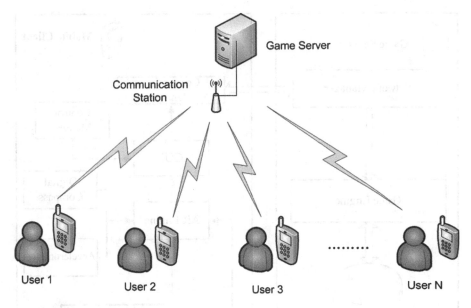

Fig. 2 The system framework

2 Design of the system Framework

The system provides service using C/S (server and client). The server provides the rendering of the virtual environment and the setting of the scenario. The client is the mobile phone platform. Because it is necessary to track the position of the users and communicate with each other, mobile phones should have 3G and GPS. The system framework is shown in Fig. 2:

The system makes use of the mobile phone platform and AR technology, which provides a virtual reconstruction of Yuanmingyuan Garden for multi-users. The main structural framework is shown in Fig. 3. The main techniques include network transmission, image recognition, model registration and so on.

3 User Interface Design

The entertainment system provides virtual modules of the reconstruction. Modules are allocated to some of target regions in Yuanmingyuan (Fig. 4). The users are divided into several groups and tasks are performed with mobile phone inside the Yuanmingyuan Garden.

First the server creates a corresponding virtual scene to the real scene of Yuanmingyuan Garden. The process is shown in Fig. 6 in detail. A visitor can download the system when entering the Yuanmingyuan Garden. The corresponding role will appear in the virtual scene. With the GPS on the cell phone, the users can

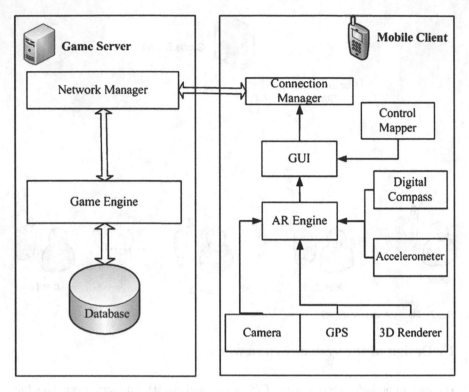

Fig. 3 Technique modules of the system

be guided to the destination. The entertainment platform will provide several tasks for the users to accomplish. With the help of GPS and virtual scene, the system can provide navigation services to guide users to accomplish the tasks. The users identify landmarks with AR technology and observe the reconstructed model of Yuanmingyuan Garden that is superimposed on the real environment as shown in Fig. 5. The users complete the corresponding tasks and get the virtual model. The game characters of the user are different. Three users make up a group to complete the task together. After collecting all of the models, the users can reconstruct Yuanmingyuan Garden together at the scene.

4 Usability Issues

Displays: For outdoor applications, the AR display should work across a wide variety of lighting conditions because we can't control the lighting to match the display. The contrast between these two conditions is huge, and most display devices cannot

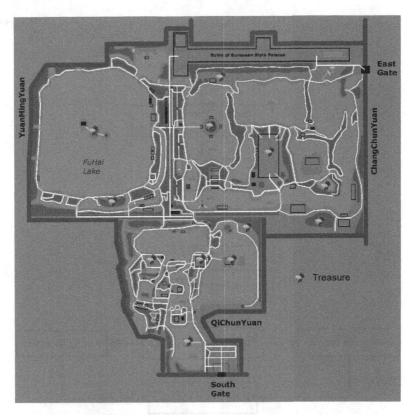

Fig. 4 Target regions of the garden map

Fig. 5 Virtual model created by AR techniques

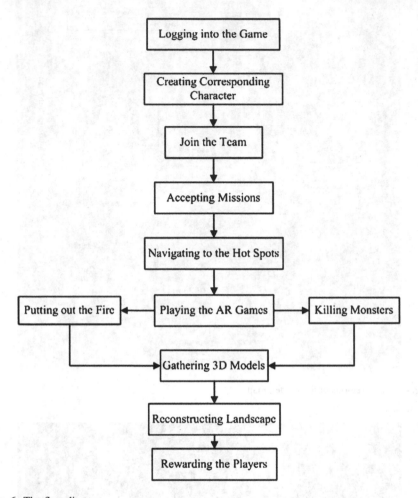

Fig. 6 The flow diagram

come close to the brightness required to match this range. It is difficult for the owner to read his display in bright sunlight.

Tracking and registration: accurate outdoor tracking is hard because of the complex surroundings. The environment is out of the user's control. And the arithmetic based image processing is very complicated.

Position Inaccuracy: The Global Positioning System provides worldwide coverage with the accuracy of about 10 meters for regular GPS and it can't provide orientation measurement. The digital compass should be used to measure orientation. But the accuracy of the compass should be also considered.

Interaction and communication: The way of the interaction and communication among the users is a problem to be solved.

5   Conclusion

There have been various researches about mobile AR on smart-phones (e.g., [4, 5, 7]) and commercial mobile AR applications have appeared for smartphones such as Layar [6].

The system proposed in this paper integrates real environment and virtual scene by the ways of entertainment and gaming in the mobile phone, which can bring unlimited entertainment to the sightseeing trip of the visitors. Through the virtual reconstruction of Yuanmingyuan Garden, it not only enables a visitor to tour its historical site but also makes him/her experience the excitements of an AR based game. The proposed system brings novel ways of entertainment to the visitors, and its users will be benefited from the learning of the cultural sites and the cultural histories. It will also be beneficial for the promotion of Yuanmingyuan Garden and the protection of cultural relics.

Acknowledgements   The project is supported by National 863 project Contact No.2008AA01Z303, 2009AA012106, National Natural Science Foundation of China Contact No. 60673198 and Fund sponsored by ZTE.

References

1. Azuma, R., Baillot, Y., Behringer, R., et al. Recent advances in augmented reality, IEEE Computer Graphics and Applications, Vo.21, No.6, pp.34–47, 2001.
2. Yetao, H., Yue, L., and Yongtian, W. AR-View: an augmented reality device for digital reconstruction of Yuangmingyuan, 8th IEEE International Symposium on Mixed and Augmented Reality 2009 - Arts, Media and Humanities (ISMAR2009). Orlando, FL, USA. IEEE. 2009:3–7.
3. Yetao, H., Yue, L., Dongdong, W., and Yongtian, W. Servocontrol Based Augmented Reality Registration Method for Digital Reconstruction of Yuanmingyuan Garden[J] Journal of Computer Research and Development. 2010.47(6):1005–1012.
4. Klein, G. and Murray, D. Parallel tracking and mapping for small AR workspaces. Proc. ISMAR 2007, 1–10. 2007.
5. Wagner, D., Langlotz, T. and Schmalstieg, D. Robust and unobtru-sive marker tracking on mobile phones. Proc. ISMAR 2008, 121–124. 2008.
6. Layar web page, 2009. http://layar.com.
7. Takacs, G., Chandrasekhar, V., Gelfand, N., Xiong, Y., and Chen, C. W. Thanos Bismpigiannis, Radek Grzeszczuk, Kari Pulli, and Bernd Girod. Outdoors augmented reality on mobile phone using loxel-based visual feature organization. In MIR '08: Proceeding of the 1st ACM international conference on Multimedia information retrieval, pp. 427–434, New York, NY, USA, 2008. ACM.

HandsOnVideo: Towards a Gesture based Mobile AR System for Remote Collaboration

Leila Alem, Franco Tecchia, and Weidong Huang

Abstract There is currently a strong need for collaborative augmented reality (AR) systems with which two or more participants interact over a distance on a mobile task involving tangible artefacts (e.g., a machine, a patient, a tool) and coordinated operations. Of interest to us is to design and develop an AR system for supporting a mobile worker involved in the maintenance/ repair of complex mining equipment. This paper presents HandsOnVideo, our on-going work towards a gesture based mobile AR system for remote collaboration. HandsOnVideo is designed and developed using a participatory design approach. Following this approach, we learnt that helpers found it natural to use their hands for pointing to objects and explaining procedures and that workers found it valuable to see the hands of the person guiding them remotely. On the other hand, we observed that this form of hand gesture interaction supported by HandsOnVideo resulted in network latency. This paper describes some of the design tradeoffs we came across during the design process and tested with representative end users. These tradeoffs will be further investigated as part of our research agenda.

1 Introduction

There are a range of real world situations in which remote expert guidance is required for a local novice to complete physical tasks. For example, in telemedicine a specialist doctor guiding remotely a non specialist doctor or nurse performing surgery on a patient [1]; in remote maintenance an expert guiding remotely a technician into repairing a piece of equipment [2]. Particularly in the field of the industrial and mineral extraction, complex technologies such as fully automated or semi automated equipments, tele-operated machines, are being introduced to improve productivity.

L. Alem (✉)
CSIRO ICT Centre, PO Box 76, Epping NSW 1710, Australia
e-mail: Leila.Alem@csiro.au

L. Alem and W. Huang (eds.), *Recent Trends of Mobile Collaborative
Augmented Reality Systems*, DOI 10.1007/978-1-4419-9845-3_11,
© Springer Science+Business Media, LLC 2011

Consequently, the maintenance and operation of these complex machines are becoming an issue. Operators/technicians rely on assistance from an expert (or more) in order to keep their machinery functioning. Personnel with such expertise, however, are not always physically located with the machine. Instead, they are often in a major metropolitan city while the technicians maintaining the machine are in rural areas where industrial plants or mine sites may be located. Therefore, there is a growing interest in the use and development of technologies to support the collaboration between a maintenance worker and a remote expert. Technologies used to support remote collaboration include email exchanges, telephone calls, video conferencing and video-mediated gesturing. Our work presented in this paper explores the potential of non-mediated hand gesture communication for remote guiding.

One of the main concerns when participants collaborate remotely is the lack of common ground for them to effectively communicate [16]. Clark and Brennan [5] define common ground as a state of mutual agreement among collaborators about what has been referred to. In the scenario of an expert guiding a novice on physical tasks, the expert speaks to the novice by first bringing attention to the object that they are going to work on. To achieve this, the referential words such as "this", "that", along with gestures such as digital annotations and hand pointing may be used. Only when the mutual understanding is built can instructions on how to perform tasks be effectively communicated. As such, attempts have been made to rebuild common ground for remote collaboration (e.g., [6, 7, 16]). Among many, shared visual space is one of the most discussed for such a purpose. A shared visual workspace is one where participants can create, see, share and manipulate artifacts within a bounded space. Real world examples include whiteboards and tabletops. Empirical studies have been conducted demonstrating the benefits of giving remote collaborators access to a shared visual space, which will be briefly reviewed in the next section.

This paper first provides a review of the literature on remote guidance systems. It then describes HandsOnVideo, the first step towards a gesture based mobile AR system for remote guiding, followed by discussions on the design tradeoffs we have encountered and a report on current progress of our work. Finally the paper concludes with a brief summary.

2 Related Work

In this section, we selectively review related work in the literature to provide background for our research.

2.1 Remote Guiding of Mobile Workers

There are many real world collaborative scenarios in which the worker is engaged in a mobile task or performing tasks on objects that are consistently moving. The mobility of the worker presents unique challenges and a few attempts have been made by researchers to address the challenges.

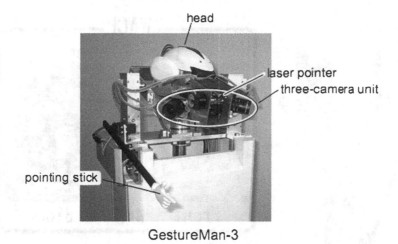

GestureMan-3

Fig. 1 Overview of GestureMan-3 system [8]

Kuzuoka et al. [8] developed a system, GestureMan-3, for supporting remote collaboration using mobile robots as communication media. See Fig. 1. The instructor controls the robot remotely and the operator receives instructions via the robot. In their system, the robot is mounted by a three-camera unit to capture the environment of the operator. It also has a laser pointer for hitting the intended position and a pointing stick for indicating the direction of the laser pointer. The movement of the robot is controlled by the instructor using a joystick.

Kurata et al. [9, 10] developed the Wearable Active Camera/Laser (WACL) system that involves the worker wearing a steerable camera/laser head. WACL allows the remote instructor not only to independently look into the worker's task space, but also to point to real objects in the task space with the laser spot. As shown in Fig. 2, the laser pointer is attached to the active camera-head and it can point a laser spot. Therefore, the instructor can observe around the worker, independently of the worker's motion, and can clearly and naturally instruct the worker in tasks.

Previous work in the area of remote guiding of mobile workers has mostly focused on supporting pointing to remote objects and/or remote area, using a projection based approach such as the laser pointing system in WACL [9, 10], or using a see through based approach such as in REAL [12] (see Fig. 3). While pointing (with a laser or a mouse) is an important aspect of guiding, research has indicated that projecting hands of the helper supports a much richer set of non verbal communication and hence is more effective for remote guiding (e.g., [7]). The next section reviews the work in this space.

2.2 Supporting Gestures in Remote Collaboration

Importance of gestures can be intuitively illustrated by hand movements that we use together with verbal and nonverbal communications in our everyday life. In fact, the

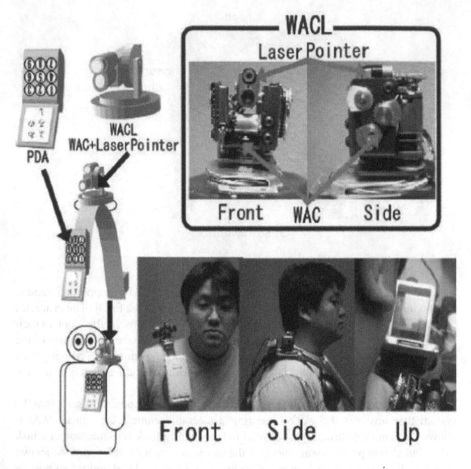

Fig. 2 The WACL system [9]

use of hand gestures in support of verbal communication is so natural that they are even used in communications when people are on the phone. Recent empirical studies have also shown that gestures play an important role in building common ground between participants in remote guiding [6].

Given that gesturing is of such importance to collaborative physical tasks a variety of systems are being developed to facilitate remote gesturing (e.g., [3, 5, 6, 7]). Most of these systems are explicitly built with the intention of enabling a remote helper (expert) to guide the actions of a local worker, allowing them to collaborate over the completion of a physical task. Results have so far suggested that such tools can increase performance speed and also improve the worker's learning of how to perform a novel task (when compared to standard video-mediated communication methods).

Fig. 3 The REAL system [12]

More specifically, Fussell et al. [6] introduced a system in which the helper can perform gestures over the video streams. In their system, gestures were instantiated as a digital form. A user study conducted by Fussell et al. demonstrated the superiority of the digital sketches over cursor pointer. More recently, Kirk et al. [7] explored the use of gesture in collaborative physical tasks using augmented reality. In particular, the guiding is supported through a mixed reality surface that aligns and integrates the ecologies of the local worker and the remote helper (Fig. 4a). The system allows the helper to see the objects in the worker's local ecology, the worker's actions on the objects in the task space, and his/her own gestures towards objects in the task space (Fig. 4b). The work of both Fussell et al. and Kirk et al. demonstrated the importance of supporting gestures. However, how gestures can be better supported with a mobile worker has not been fully understood.

3 The System of HandsOnVideo

Our literature review suggested the following requirements for AR remote guiding systems in industry.

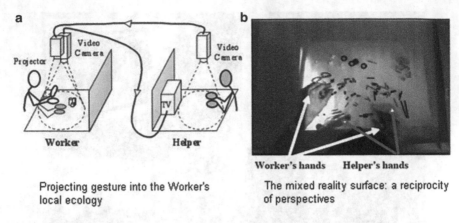

Fig. 4 Projected hands [7]

- The need to support the mobility aspect of the task performed by the worker using wearable computers and wearable cameras.
- The need to allow helpers to guide remotely using their hands in order to not only provide reference to remote objects and places, but also support procedural instructions.

These requirements are consistent with our observations of maintenance workers and our understanding of their needs. In this section, we introduce our HandsOnVideo system that is developed to address the above needs. In particular, HandsOnVideo captures the hand gestures of the helper and projects them onto a near eye display worn by the worker. The system is composed of 1) a helper user interface used to guide the worker remotely using a touch screen device and an audio link, and 2) a mobile worker system composed on a wearable computer, a camera mounted on a helmet and a near eye display (a small device with two screens); see Fig. 5. More details of the design of our remote guiding system and its technical platform are described in the following subsections.

3.1 Worker Interface Design

When it comes to display information to a mobile worker, there are a range of displays that can be chosen from for this purpose, including hand-held displays, wrist-worn displays, shoulder-worn displays and head-worn displays. Since we aimed to develop a system that can be used in mine sites, and the environment in mine sites can be noisy, dusty and unpredictable, we decided to configure our own worker interface that makes most use of worker's outfits and is less dependent on the environment.

Fig. 5 Worker interface

Workers usually wear helmets while working in mining sites for safety reasons. We therefore made use of the helmet and attached a near-eye display under the helmet. As shown in Fig. 5, the near-eye display is light, easy to put on and comfortable to wear, compared to other types of head-worn displays such as optical or video see-through displays. The worker can easily look up and see video instructions shown on the two small screens, and at the same time he/she can see the workspace in front of him/her with little constraint. We also tested the display with real users. The feedback from them during the design process was very positive with the near-eye display.

3.2 *Helper Interface Design*

We adopted a participatory approach for the design of the helper interface. Our aim was to come up with a design that fulfils the users' needs and that is as intuitive to use as possible. Our initial step consisted of observing maintenance workers and developing a set of requirements for the helper user interface (UI) based on our understanding of their needs.

- The need for supporting complex hand movements such as: "take this and put it here", "grab this object with this hand", and "do this specific rocking movement with a spanner in the other hand".
- Mobility of the worker during the task, as they move from being in front of the machine to a tool area where they access tools, to the back of the machine to check valves etc.

Fig. 6 Maintenance and assembly task

- The helper may need to point/gesture in an area outside the field of view of the worker. Therefore there is a need to provide the helper with a panoramic view of the remote workspace.

We then designed a first sketch of the interface consisting of a panoramic view of the workspace and a video of the worker's view. The video provides a shared visual space between the helper and the worker that is used by the helper for pointing and gesturing with their hands (using unmediated gesture). This shared visual space augmented by the helper's gestures is displayed real time on the near eye display of the worker (image + gestures).

The helper UI consists of:

- A shared visual space which displays, by default, the video stream captured by the remote worker's camera. This space occupies the central area of the touch table.
- A panoramic view of the worker's workspace which the helper can use for maintaining an overall awareness of the workspace. This view can also be used by the helper for bringing the worker to an area that is outside his/her current field of view. The panoramic view occupied the lower end of the touch table.
- Four storage areas, two on each side of the shared visual space, to allow the helper to save a copy of a particular scene of the workspace and reuse it at a later stage of the collaboration.

We performed four design iterations of our UI, testing and validating each design with a set of representative end users on the following three maintenance/repair tasks (Fig. 6):

- Repairing a photocopy machine
- Removing a card from a computer mother board and
- Assembling Lego toys

Fig. 7 The helper control console (*left*) and the worker wearable unit (*right*)

Over 12 people have used and trialled our system, providing valuable feedback on how to improve the helper UI and more specifically the interactive aspect of the UI: the selection of a view, the changing of the view in the shared visual space and the storage of a view. The aim was to perform these operations in a consistent and intuitive manner, for ease of use. The overall response from our representative end users pool is that our system is quite intuitive and easy to use. No discomfort has been reported to date with the near eye display of the worker system.

3.3 Technical Platform of HandsOnVideo

Our platform draws on previous experience in the making of the REAL system [12], a commercial, wearable, low-power augmented reality system that employs an optical see-through visor (LiteEye 750). REAL has been used for remote maintenance in industrial scenarios. In particular, HandsOnVideo makes use of the XVR platform [14], a flexible, general-purpose framework for VR and AR development. The architecture of our system is organized around two main computing components: a worker wearable device and a helper station, as seen in Fig. 7.

Wearable computers have usually lower computing capability in comparison to desktop computers. To take into account the usual shortcomings of these platforms all our software has been developed using an Intel Atom N450 as a target CPU (running Microsoft Windows XP). It presents reasonable heat dissipation requirement and peak power consumptions below 12 watts, easily allowing for battery operation. A Vuzix Wrap 920 HMD mounted on a safety helmet was used as the main display of the system. The arrangement of the display is such that the upper part of the worker field of view is occupied by the HMD screen so that the worker can look at the screens by just looking up, while the lower part remains non occluded. With such an arrangement, what is displayed on the HMD gets used as a reference, but then the worker performs

Fig. 8 Layout of the helper screen

all his/her actions by directly looking at the objects in front of him/her. CMOS USB camera (Microsoft Lifecam HD) is mounted on top of the worker's helmet (as seen in Fig. 5) and allows the helper to see what the worker is doing with his/her hands. A headset is used for the worker-helper audio communication.

The main function of the wearable computer is to capture the live audio and video streams, compress them in order to allow network streaming at a reasonable low bit rate, and finally deal with typical network related issues like packet loss and jitter compensation. To minimize latency we use a low level communication protocol based on UDP packets, data redundancy and forward error correction, giving us the ability to simulate an arbitrary values of compression/decompression/network latency, with a minimum measured value around 100 ms. Google's VP8 video compressor [15] is used for video encoding/decoding, and the Open Source SPEEX library is used for audio, with a sampling rate of 8Khz. Please note that at the same time the wearable computer also acts as a video/audio decoder, as it receives live streams from the helper station and renders them to the local worker.

The main component of the helper station is a large (44 inches) touch-enabled display, driven by a NVidia GeForce graphic card mounted on a Dual Core 2.0 Ghz Intel workstation (Windows XP). The full surface of the screen is used as a touch-enabled interface, as depicted in Fig. 8.

Occupying the central portion of the screen is an area that shows the video stream captured by the remote worker camera: it is on this area that the gesture of the helper takes place. On the side of the live stream there are 4 slots, initially empty, where at any moment it is possible to copy the current image of the stream. This can be useful to store images of particular importance for the collaborative task, or snapshots of locations/objects that are recurrent in the workspace. Another high-resolution webcam (Microsoft Lifecam HD) is mounted on a fixed support attached to the frame of the screen, and positioned to capture the area on the screen where the video stream is displayed (see

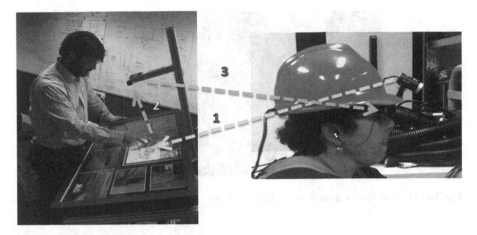

Fig. 9 Data capture and display

Fig. 9): the camera captures what is shown on the touch screen (see arrow 1) and the hand gestures performed by the helper over that area too (see arrow 2).The resulting composition (original image plus the hand gesture on top) is once again compressed and streamed to the remote worker, to be displayed on the HMD (see arrow 3). The overall flow of information is represented in the diagram of Fig. 9.

4 Discussions and Current Progress

Designing a useable remote guiding system requires a close involvement of end users, and the ability to capture and address the interaction issues they raise while using the system. This is not an easy task and during the system design and development process, we have encountered a number of challenges. These challenges include:

- The trade off between the richness of the gesture supported by the system and the resulting latency it introduces. We are currently exploring means by which we can extract the hands of the helper from the shared visual space and display the hands on the local video view of the worker.
- The trade off between the quality of the image/video projected and network latency. We are currently exploring ways in which we can provide a high resolution video of a subset of the shared visual space.
- The trade off between supporting the mobility of the worker while maintaining spatial coherence. Because of the worker's mobility these are sometimes discrepancies between the view projected in the worker's display and the view the worker has of his physical workspace. This may disorient workers. There is a need for workers to maintain a spatial coherence. We are currently exploring gesture based interactions to allow the worker to change the view displayed on their view. As the worker moves around, the shared visual view changes, gesturing

Fig. 10 The local view augmented with helper's hands

on a changing target could become challenging. To address this issue, gesturing (pointing to a location or an object and showing orientation and shape, etc.) was initially performed on a still image. Helpers were required to freeze the video view in order to gesture. Gesturing on a still image created an extra workload for the helper and results in the shared visual space not synchronised with the view of the physical workspace. We are currently investigating how to allow the helper to gesture on the video view.

We have conducted a usability testing of our system [17], and are currently improving some technical features which we expect to make HandsOnVideo an AR enabled system. In particular, we are developing a hand-extraction algorithm that captures and extracts the helper's hands from the shared visual space. The system compresses the hands (without the background) and uses the resulting images to achieve a chroma-key overlapping of the helper hands on the local copy of the images captured by the worker video (see Fig. 10). It is expected that the hand extraction algorithm will significantly decrease the system bandwidth requirement, and hence greatly improving the quality of the images displayed at the worker end. Another advantage of this algorithm is that it will allow us to differentiate the helper hands from the worker hands by for example displaying the helper's hand in grey shade. This differentiation should improve user's experience as some end users reported being confused at time.

5 Conclusion

We have reviewed, in this paper, the literature in augmented reality remote guiding, and put forward the case for supporting richness of gesture and mobility of the worker. We described HandsOnVideo, the on-going work towards a mobile AR system for remote collaboration. The system was designed using a participatory design approach. Our key research drive was to develop a remote guiding system that is truly useful, enjoyable, easy to use, reliable effective and comfortable for end users. The design approach we have taken has allowed us to test and trial, with end users, a number of design ideas. It also enabled us to understand from a user's perspective some of the design tradeoffs. We plan to explore these tradeoffs in a series of laboratory experiments and believe that exploring these tradeoffs will provide a solid basis on how to design useful remote guiding systems.

Acknowledgements We would like to thank the Future mine research theme of the CSIRO Minerals Down Under flagship for sponsoring and supporting this research effort.

References

1. Palmer, D., Adcock, M., Smith, J., Hutchins, M., Gunn, C., Stevenson, D., and Taylor, K. (2007) Annotating with light for remote guidance. In *Proceedings of the 19th Australasian Conference on Computer-Human interaction: Entertaining User interfaces* (Adelaide, Australia, November 28 - 30, 2007). OZCHI '07, vol. 251. ACM, New York, NY, 103–110.
2. Kraut, R. E., Miller, M. D. and Siegel, J. (1996) Collaboration in performance of physical tasks: effects on outcomes and communication. CSCW '96: *Proceedings of the 1996 ACM conference on Computer supported cooperative work*, ACM, 1996, 57–66.
3. Kraut, R. E., Fussell, S. R., and Siegel, J. (2003) Visual information as a conversational resource in collaborative physical tasks. *Hum.-Comput. Interact.* 18, 1 (Jun. 2003), 13–49.
4. Li, J., Wessels, A., Alem, L., and Stitzlein, C. (2007) Exploring interface with representation of gesture for remote collaboration. In *Proceedings of the 19th Australasian Conference on Computer-Human interaction: Entertaining User interfaces* (Adelaide, Australia, November 28 - 30, 2007). OZCHI '07, vol. 251. ACM, New York, NY, 179–182.
5. Clark, H. H. and Brennan (1991) Grounding in communication. *Perspectives on socially shared cognition*. American Psychological Association, 1991.
6. Fussell, S. R., Setlock, L. D., Yang, J., Ou, J., Mauer, E., and Kramer, A. D. I. (2004) Gestures over video streams to support remote collaboration on physical tasks. *Hum.-Comput. Interact.*, L. Erlbaum Associates Inc., 19, 273–309.
7. Kirk, D. and Stanton Fraser, D. (2006) Comparing remote gesture technologies for supporting collaborative physical tasks. CHI '06: Proceedings of *the SIGCHI conference on Human Factors in computing systems*, ACM, 1191–1200.
8. Kuzuoka, H., Kosaka, J., Yamazaki, K., Suga, Y., Yamazaki, A., Luff, P., and Heath, C. (2004) Mediating dual ecologies. CSCW '04: Proceedings of *the 2004 ACM conference on Computer supported cooperative work*, ACM, 477–486.
9. Kurata, T., Sakata, N., Kourogi, M., Kuzuoka, H., and Billinghurst, M. (2004) Remote collaboration using a shoulder-worn active camera/laser. *Eighth International Symposium on Wearable Computers* (ISWC'04), 1, 62–69.
10. Sakata, N., Kurata, T., Kato, T., Kourogi, M., and Kuzuoka, H. (2003) WACL: supporting telecommunications using - wearable active camera with laser pointer Wearable Computers, 2003. Proceedings. *Seventh IEEE International Symposium on*, 2003, 53–56.
11. Langlotz, T., Wagner, D., Mulloni, A., and Schmalstieg, D. (2010) Online Creation of Panoramic Augmented Reality Annotations on Mobile Phones. *IEEE Pervasive Computing*, 10 Aug. 2010. IEEE computer Society Digital Library. IEEE Computer Society.
12. R.E.A.L. (REmote Assistance for Lines), (c)(TM) SIDEL S.p.a. & VRMedia S.r.l, http://www.vrmedia.it/Real.htm.
13. Lapkin et al. (2009) *Hype Cycle for Context-Aware Computing*. Gartner research report, 23 July 2009. ID Number: G00168774.

14. Carrozzino M., Tecchia F., Bacinelli S., and Bergamasco M. (2005) Lowering the Development Time of Multimodal Interactive Application: The Real-life Experience of the XVR project, Proceedings of *ACM SIGCHI International Conference on Advances in Computer Entertainment Technology* (ACE'05), Valencia, Spain.
15. The Google WebM project, http://www.webmproject.org/, Google Inc.
16. Ranjan, A., Birnholtz, J. P., and Balakrishnan, R. (2007) Dynamic shared visual spaces: experimenting with automatic camera control in a remote repair task. CHI'07: Proceedings of *the SIGCHI conference on Human factors in computing systems*, ACM, 1177–1186.
17. Huang, W. and Alem, L. (2011) Supporting Hand Gestures in Mobile Remote Collaboration: A Usability Evaluation. In *Proceedings of the 25th BCS Conference on Human Computer Interaction*, 4 July – 8th July, 2011, in Newcastle-upon-Tyne, UK.

Dynamic, Abstract Representations of Audio in a Mobile Augmented Reality Conferencing System

Sean White and Steven Feiner

Abstract We describe a wearable audio conferencing and information presentation system that represents individual participants and audio elements through dynamic, visual abstractions, presented on a tracked, see-through head-worn display. Our interest is in communication spaces, annotation, and data that are represented by auditory media with synchronistic or synesthetic visualizations. Representations can transition between different spatial modalities as audio elements enter and exit the wearer's physical presence. In this chapter, we discuss the user interface and infrastructure, SoundSight, which uses the Skype Internet telephony API to support wireless conferencing, and describe our early experience using the system.

1 Introduction

The proliferation of mobile phones and voice over IP (VoIP) applications such as Skype enables audio communication throughout many aspects of modern life. Conversations are no longer limited to a few individuals, but may encompass large groups from diverse locations and sources. At the same time, portable music players, mobile phones, and hand-held computers provide easy access to sound generation in every environment. Projects such as Freesound [1] provide widely available geospatially tagged sounds from their websites. This combination of technologies— pervasive audio communication and geocoded sound—provides new opportunities for overlaying and interacting with audio in the environment.

In the context of our research into mobile, augmented reality electronic field guides (EFG) for botanical species identification [23], we have been developing

S. White (✉)
Columbia University & Nokia Research Center, 2400 Broadway,
Suite D500, Santa Monica, CA 90404
e-mail: sean.white@nokia.com

L. Alem and W. Huang (eds.), *Recent Trends of Mobile Collaborative Augmented Reality Systems*, DOI 10.1007/978-1-4419-9845-3_12, © Springer Science+Business Media, LLC 2011

ways in which audio from multiple applications can be represented and presented in the user interface. Our interest is in supporting communication spaces, audio annotations, and audio data, together with the visual interactions that already exist in our prototype EFG.

For example, a botanist in the field must effectively communicate with remote experts, such as colleagues that may be over the next hill or on another continent, to discuss new specimens. They will seek to discover, listen to, or create audio annotations, which may be tethered to a geographic location, a concept (such as a species), or even a mobile/movable object (such as another field botanist or a physical specimen). While audio provides an easy and familiar media channel for these interactions, difficulties arise when using it in isolation. Multiple speakers may be hard to disambiguate and annotations may be hard to localize.

We seek to preserve the benefits of audio, such as low attentional demand and light-weight verbal communication, while improving the efficacy of audio in augmented reality, by developing an increased sense of presence from audio sources, the ability to distinguish sources using spatial models, and localization of specific sound sources. In doing so, we intend to support cognitive processes such as spatial memory, peripheral discovery, and analytic reasoning about auditory information.

Direct spatialization of audio provides one means of addressing the difficulties in displaying audio sources [22]. This can be done either through manipulation of the sound or through recording and playback techniques such as those presented by Li et al. [10], who use microphone arrays and speaker arrays to acquire and reproduce 3D sound. However, this requires additional computation and bandwidth, and such audio-only spatial cues are not always effective at sound separation.

As an adjunct and alternative, we have been investigating the use of spatialized, synesthetic visual representations of sound for communication and annotation, which we discuss in the remainder of this chapter. We start by presenting research related to synesthesia and representation, audio communication spaces, and visual representation of audio. Next, we introduce our user interface and system, SoundSight (Figure 1), and discuss visual representations, spatial frames of presentation, and transitions across spaces. We describe scenarios for collaborative communication and annotation to explore use of the system. Then, we discuss user interface prototypes that provide a testbed for novel techniques. We end with observations about the use of the prototypes, conclusions, and a discussion of our ongoing and future work.

2 Background

This research draws from two main areas: visual representation of audio and spatial conferencing systems. Discussions of visual representation of audio often use the term *synesthesia*, whose Greek roots mean to perceive together. This term refers to the phenomenon of stimulation in one sensory modality giving rise to sensation in another. We primarily consider sound and vision, although our interest is in the full set of senses.

Fig. 1 Frames from a video made using SoundSight. (Top left) Two body-fixed audio elements (green and red), as seen by the local conference participant. (Top right) A remote participant (represented by the red element), using Skype over a Bluetooth headset, enters the room. (Bottom left) The formerly remote participant moves into video tracking range near her audio element. (Bottom right) Her audio element transitions to object-fixed representation (relative to her) when her badge is recognized and tracked

Van Campen [4] provides a historical review of artistic and psychological experiments in synesthesia. He discusses the evolution of contrasting theories of the sensory and neurological causes of the phenomenon, such as independent and unified sensory apparatuses. More recently, Plouznikoff et al. [15] proposed using artificial synesthesia to improve short term memory recall and visual information search times.

Snibbe and Levin [21] presented artistic experiments in dynamic abstractions, with the stated goal of creating phenomenological interfaces that directly engage the unconscious mind. In a more instrumental form, Pedersen and Sokoler [14] developed AROMA to explore abstract representations of presence. Their primary aim was to remap signals to support purely abstract visual forms of peripheral awareness. They observed that abstract representations provide a better, non–attention-demanding awareness than media-rich ones. No actual audio from the source is presented and their abstract forms have no specific spatial location. In contrast, our system combines abstract representations, placed in spatial locations, together with actual audio sources. Laurel and Strickland [9] developed the concept of voiceholders in their multi-person PlaceHolder VR. In their system, voiceholders were identical 3D objects that could be touched to record or play back audio stories. While they did not directly represent motion of the sound, the objects could be placed spatially. We build on these ideas and use unique and dynamic representations of each audio source.

A variety of communication systems have explored spatial representations and cues. Hydra [19] used physical elements to provide spatial cues for videoconferencing. Benford et al. [2] introduced a spatial model of interaction for supporting social communication in distributed virtual environments. Singer et al. [20] developed mixed audio communication spaces with a variety of user interfaces, including Vizwire, using a visual spatial metaphor, and Toontown, based on active objects. They also discussed privacy issues. While not intended specifically for representing audio elements, FieldNote [13] recorded notes with contextual information in the field that could be associated with a location. GeoNotes [5] focused on the social aspect of creating and interacting with geocoded information. Sawhney and Schmandt [17] found that spatial cues aided auditory memory in their Nomadic Radio system, increasing the number of simultaneous audio streams a user could comprehend. Schmandt also found that spatial characteristics facilitate navigation and recall in the Audio Hallway [18]. Rather than focus on the spatialization of the audio source, as these systems do, we focus on visual cues that create a spatial perception of the audio source.

Rodenstein and Donath's Talking in Circles [16] supports shared-space multimodal audio conferences, emphasizing the 2D arrangement of speakers and graphics synchronized to speakers. Their primary concern is identity, presence, and activity. 2D spatial relationships reflected mutual attenuation represented in a third-person overview. In contrast, our research investigates 3D representations, a 3D space for positioning audio, and a first-person perspective on the audio. Billinghurst et al. [3] developed a wearable spatial conferencing system that spatialized static images of speakers in an audio conference, providing cues for localizing and identifying participants. While inspired by this work, we look at dynamic, abstract visual representations that do not require concrete visual representation of speakers or sound sources.

3   SoundSight

SoundSight helps the user "see" sounds. The system provides visual representations for audio elements such as audio conference participants, voicemail, music sources, and audio annotations. Audio elements come in many forms. They can be live or pre-recorded; actual and unmediated or virtual; monaural or spatialized in simple stereo, binaural, or 3D.

Each sound or group of sounds is embodied by a 3D visual representation. The representation may be associated with a single sound source or a combination of sounds. Thus, a set of four groups of four people can be represented as 16 individual representations or as four group representations. The dynamics of each representation, discussed in the next section, represent either individual or group

Fig. 2 The shape of the representation changes with the audio source. In this case, a cube changes size in reference to a sphere. From top left to bottom right, the cube increases in size with the volume of the audio source

audio, respectively. The spatial layout of visual representations is presented to the viewer based on a combination of source location and listener preference.

3.1 Representation

Visual representation of the audio source provides a means of reflecting presence for discovery and awareness, localizing elements in space, and observing consistent identity. Although we do not require the representation to be abstract, we posit that an abstract representation may support different interactions than a photorealistic one. McCloud [11] suggests that more abstract visual representations encourage the viewer to project their own cognitive models on the representation to help fill in gaps in story, narrative, or description. McLuhan [12] also argued for a distinction between hot and cold media, although he was less concerned with specific representations within a given medium.

SoundSight supports a continuum of representations that vary from completely abstract to photorealistic. Representations can change based on the stage of a conversation or the requirements of the listener. For example, an audio conference may start with an image of a remote participant, but change to a dynamic, abstraction once the conversation has been initiated. This lets the user adjust the amount of visual attention required by the system, while still benefiting from visual representations.

All representations are considered to be dynamic, in that they will reflect the dynamics of the audio stream in the visual domain. We do this because the perception of action is closely tied to presence [14]. We note that we initially represented sound levels with single objects changing in scale or rotation. However, we found that this did not provide any constant visual reference to the normal state. We addressed this by having geometric objects change in size relative to existing geometric objects (Figure 2).

3.2   Spatial Presentation

Elements may be presented in a variety of ways, based on whether the location veridically represents a sound source in the world, abstractly represents the location of the source, or is located based on the spatial requirements of the listener. The listener may also need to transition audio elements from one spatial presentation to another. We consider an egocentric model of presentation relative to the listener. We define *locus* to mean the collection of points sharing a common property, and *spatial locus* to represent a specific coordinate space or stabilization space. Feiner et al. [6] use world-fixed, surround-fixed, and display-fixed to describe the spatial loci within which window elements are displayed to a user. Their surround is a virtual sphere centered around the user. Billinghurst et al. [3] describe three models of spatial presentation: world-stabilized, body-stabilized, and head-stabilized. These represent information fixed to real world locations, relative to orientation of the body, or relative to the user's viewpoint, respectively.

We build on these models, further distinguishing between variations in body loci and between the world and objects in the world as loci. Therefore, we consider five spatial loci:

World-fixed. Elements are fixed to world coordinates external to the user. Their locations are not expected to change over time.

Object-fixed. Elements are presented fixed relative to a stationary or movable object that may be external to or carried by the user. While this has also been referred to as world-fixed [6], we use the term object-fixed to further distinguish spatial loci that may be moving within the world.

Body-fixed. Elements are presented fixed relative to the torso. Head orientation relative to torso orientation provides simple navigation in the space, while maintaining the mobility and spatial consistency of the perceptual objects in the space. When speaking with a group, we often orient our body to face the centroid of the group and then use head-movement to shift attention to individuals in the group. Certainly some body shifting occurs, but the simpler act of head-movement predominates.

Compass-fixed. Elements are presented relative to head orientation, but independent of position. For instance, a botanist might have colleagues in a conference call who are west of her and always represented in that direction. Auditory cues for wayfinding may also be presented in this locus.

Display-fixed. Elements are presented fixed relative to the visual display. The display does not explicitly change based on orientation or position, although it might change based on context or location. For clarity, we precede the term by an adjective denoting the specific display; for example, head-worn-display–fixed or tablet-display–fixed.

We use these distinctions as shorthand for discussing presentation in the user interface. Representational elements can be presented to the user in these spatial loci, either in isolation or composited.

Fig. 3 (top left) Visual representation presented object-fixed to the remote speaker. When the speaker is no longer in view, the visual representation can transition to other loci such as (top right) display-fixed to the desktop application, (bottom left) body-fixed from the listener's point of view, and (bottom right) object-fixed to a local object

3.3 Transitions

Just as we make transitions in the actual world as we move from place to place, elements in SoundSight can transition from one spatial locus to another. For example, user A may be speaking to remote user B, who then enters and later leaves A's physical presence during the conversation. As shown in Figure 1, SoundsSight supports the transitions by changing the spatial locus of B from body-fixed (relative to A's body) to object-fixed relative to B (who is wearing a tracked badge), and then back to body-fixed when B leaves. Figure 3(a) shows a physically present speaker and Figure 3(b–c) shows different transitions of the representation when the speaker leaves the immediate vicinity.

In another example, a user may discover a sound element as they are walking along a path. They can stay and listen to the object or they can carry it with them and have it become display-fixed or body-fixed.

In each case, the transition is from one spatial locus to another, but the form of transition is relatively undefined. We consider three types of transitions, although there are certainly others that exist. The first transition is a simple warp, the second fades the element out of one locus and into another, and the third animates between the two loci.

Of primary importance is the association of the element from one space as it moves to another. This is reflected by maintaining the visual representation across spaces. Elements may also transition by copying, rather than moving, to maintain the cognitive model that the original sound source is still located in a particular position.

4 System Design and Implementation

4.1 Hardware

Our system uses a Sony U750 hand-held and Lenovo ThinkPad X41 tablet PC running Windows XP. The computer is connected to a Sony LDI-D100B 800×600 resolution, color, see-through head-worn display, two orientation trackers, and a video camera. The head is tracked by an InterSense InertiaCube3 hybrid inertial orientation tracker and the waist is tracked by a similar InterSense InertiaCube2. Audio is played through Sennheiser stereo headphones. A Unibrain Fire-I board color video camera is mounted on the head-worn display and used to track objects in the world. Additional portable power for the Fire-I is provided using a Tri-M PC104 HE104 supply. Remote participants use a phone, networked PC with audio, or laptop with Bluetooth headset.

4.2 Software Architecture

The software architecture consists of modules for tracking, visual display, auditory display, analysis/synthesis of audio, and network transport of data, all developed in C++.

The tracking module maintains models of the torso and head trackers, as well as object and world locations from the video camera. The orientation trackers are accessed using the InterSense isense.dll and the fiducials used to find objects and the world are located using ARToolkit.

The visual display is rendered using OpenGL. Individual objects are managed and located within each spatial locus. Compositing of loci is done within this module.

The auditory display module is used to manage live local sounds from the microphone, pre-recorded sounds, and audio conferencing. We use the BASS audio library to manage local audio, and Skype and the Skype API to manage audio conferencing.

The synthesis module, also implemented using BASS, analyzes the data to extract features such as amplitude and frequency distribution from the signal, which are then synthesized into a smaller data stream. The synthesized information is then sent through the transport module along with the actual data stream.

Dynamic display requires knowledge about the individual sound sources. One way to acquire this information is to access each of the individual streams on the listener side. Since only the combined stream is available through Skype, the individual sound streams must first be processed at the source for amplitude and frequency data and this information then passed to the receiver to complement the undifferentiated Skype audio stream. We developed our own UDP-based client/server application to analyze sound locally and then send the data to the listener alongside the Skype audio data.

5 Initial User Experience

In an informal pilot, we explored audio conferencing similar to the scenarios discussed earlier. Our goal was an exploratory study to understand basic usability, participants' perception of presence, and participants' ability to distinguish different sound sources.

Five participants (four male, one female), ages 20–28, were recruited via word of mouth. Participants received no compensation for their time. We used the prototype to present a single monaural stream into which three separate monaural audio streams of speech were mixed. In the *audio-only* condition, participants wore only headphones. In both the *static-visual* and *dynamic-visual* conditions, participants wore a head-worn display in addition to headphones, and visual representations were presented. In all conditions, the three audio elements were presented body-fixed. In the static-visual condition, the visual representations were static and did not change shape with the audio, while in the *dynamic-visual* condition, the each visual representation changed dynamically with the audio source.

5.1 Task

Participants were first given the opportunity to communicate with a remote individual using the three different conditions in an open format to become familiar with the experience. After initial use of the system, participants were presented with multiple audio sources in each of the different conditions and asked to report on their ability to distinguish individual sound sources and perception of presence.

5.2 Results

Synesthetic, dynamic changes in the visual representation helped identify a given object as the source of the audio. Users talked about seeing the rhythm of the audio in the movement of the visual representation and knowing it was present.

This phenomenon is referred to as the *ventriloquism effect*[1] in psychological and perceptual literature.

We observed that elements that cannot be disambiguated in the audio-only condition are easily separated in the dynamic-visual condition. In our experiments with three sources, which did not use spatialized audio, users were able to discriminate amongst the audio sources and localize each source very quickly based on visual cues. During exit interviews, participants told us that they assumed that the sources were spatialized or presented in stereo and were surprised to learn that the sounds were monaural.

The dynamic-visual representations were perceived as more present than audio-only or static-visual representations, but too attentionally demanding in some cases. We addressed this by changing the size and shape of the representations. During movement and motion, the preferred location of larger audio elements was to the sides of the visual display. We believe this may be addressed with smaller, transparent, less obtrusive representations or more specific locations based on awareness of the actual background [8].

We note that certain aspects of face-to-face and video conferencing, such as gaze, are not present here. While we recognize gaze as important for collaboration, we consider it one of many potential cues that may or may not be present in a communication space.

Although objects changed dynamically based on sound levels, their location in a given locus was rigid. This made some audio sources appear closely associated when they were not. Small independent, animated changes in location may reinforce the individuality of elements.

6 Conclusions and Future Work

In this chapter, we have presented a preliminary description of SoundSight, a system for dynamic, abstract audio representations in mobile augmented reality. We discussed user interface techniques for representations, spatial loci, and transitions between loci, and described our underlying implementation. Finally, we described initial informal user experience with the system, and observed that spatialized, dynamic, visual representations of audio appear to increase the sense of presence and support localization and disambiguation of audio sources.

We are interested in continuing this line of research by developing systems for handheld augmented reality and expanding to incorporate more types of audio elements. In the process, we will be exploring infrastructure for creating, finding, and collecting audio elements in the field to create a unified representation of the audio space. We will extend this to sharing collected material among users.

---

[1] The "ventriloquism effect" or "ventriloquist effect" refers to the perception of speech or auditory stimuli as coming from a different location than the true source of the sound, due to the influence of visual stimuli associated with the perceived location.

Acknowledgements We would like to thank Bill Campbell for his help with the Skype API; Dominic Marino, Randall Li, Michael Wasserman, and members of the CGUI lab for their discussion of abstract representations; and Lauren Wilcox for her all her help with video. This work was funded in part by NSF Grant IIS-03-25867 and a generous gift from Microsoft Research.

References

1. Freesound - http://freesound.iua.upf.edu/
2. S. Benford, J. Bowers, L.E. Fahlen, J. Mariani, and T. Rodden, "Supporting Cooperative Work in Virtual Environments," *The Computer Journal*, vol. 37, 8, pp. 653–668, 1994.
3. M. Billinghurst, J. Bowskill, M. Jessop, and J. Morphett, "A wearable spatial conferencing space," *Proc. ISWC*, Pittsburgh, PA, pp. 76–83, 1998.
4. C. van Campen, "Artistic and Psychological Experiments with Synesthesia," *Leonardo*, vol. 32, 1, pp. 9–14, 1999.
5. F. Espinoza, P. Persson, A. Sandin, H. Nystrom, E. Cacciatore, and M. Bylund, "GeoNotes: Social and Navigational Aspects of Location-Based Information Systems," *Proc. Ubicomp*, Atlanta, GA, pp. 2–17, 2001.
6. S. Feiner, B. MacIntyre, M. Haupt, and E. Solomon, "Windows on the world: 2D windows for 3D augmented reality." *Proc. UIST 1993*, Atlanta, Georgia, United States: ACM Press 1993, pp. 145–155.
7. L.N. Foner, "Artificial synesthesia via sonification: A wearable augmented sensory system," *Proc. ISWC*, Cambridge, MA, pp. 156–157, 1997.
8. J.L. Gabbard, J.E. Swan, D. Hix, R.S. Schulman, J. Lucas, and D. Gupta, "An empirical user-based study of text drawing styles and outdoor background textures for augmented reality," *Proc. IEEE VR*, pp. 11–18, 2005.
9. B. Laurel and R. Strickland, "PLACEHOLDER: Landscape and narrative in virtual environments," *Proc. ACM Multimedia*, pp. 121–127, 1994.
10. Z. Li, R. Duraiswami, and L.S. Davis, "Recording and reproducing high order surround auditory scenes for mixed and augmented reality," *Proc. IEEE ISMAR*, Arlington, VA, pp. 240–249, 2004.
11. S. McCloud, *Understanding Comics: The Invisible Art*. New York: Harper Collins, 1994.
12. M. McLuhan, *Understanding Media: The Extensions of Man*. Cambridge, MA: The MIT Press, 1964.
13. J. Pascoe, N. Ryan, and D. Morse, "Using While Moving: HCI Issues in Fieldwork Environments," *ACM Trans. on Computer Human Interaction*, vol. 7, 3, pp. 417–437, 2000.
14. E. Pedersen and T. Sokoler, "AROMA: Abstract representation of presence supporting mutual awareness," *Proc. CHI*, pp. 51–58, 1997.
15. N. Plouznikoff, A. Plouznikoff, and J.M. Robert, "Artificial grapheme-color synesthesia for wearable task support," *Proc. IEEE ISWC*, pp. 108–111, 2005.
16. R. Rodenstein and J. Donath, "Talking in circles: Designing a spatially-grounded audioconferencing environment," *Proc. CHI*, The Hague, The Netherlands, pp. 81–88, 2000.
17. N. Sawhney and C. Schmandt, "Design of Spatialized Audio in Nomadic Environments," *Proc. Int. Conf. on Auditory Display*, CA, 1997.
18. C. Schmandt, "Audio hallway: A virtual acoustic environment for browsing," *Proc. UIST*, San Francisco, CA, pp. 163–170, 1998.
19. A. Sellen and B. Buxton, "Using Spatial Cues to Improve Videoconferencing," *Proc. CHI*, pp. 651–652, 1992.
20. A. Singer, D. Hindus, L. Stifelman, and S. White, "Tangible progress: less is more in Somewire audio spaces," *Proc. CHI*, Pittsburgh, PA, pp. 104–111 1999.
21. S.S. Snibbe and G. Levin, "Interactive dynamic abstraction," *Proc. 1st Int. Symp. on Non-photorealistic Animation and Rendering*," Annecy, France pp. 21–29, 2000.

22. E.M. Wenzel, "Localization in Virtual Acoustic Environments," *PRESENCE*, vol. 1, 1, pp. 80–107, 1992.
23. S. White, S. Feiner, and J. Kopylec, "Virtual Vouchers: Prototyping a Mobile Augmented Reality User Interface for Botanical Species Identification," *Proc. IEEE 3DUI*, Alexandria, VA, pp. 119–126, 2006.

Facilitating Collaboration with Laser Projector-Based Spatial Augmented Reality in Industrial Applications

Jianlong Zhou, Ivan Lee, Bruce H. Thomas, Andrew Sansome, and Roland Menassa

Abstract Spatial Augmented Reality (SAR) superimposes computer generated virtual objects directly on physical objects' surfaces. This enables user to interact with real world objects in a natural manner. This chapter investigates SAR techniques and summarizes advantages with the difficulties of SAR in presenting digital information to users. The chapter then presents a concept of portable collaborative SAR. The concept utilizes both projector-based SAR and Head-Mounted-Display (HMD) based Augmented Reality (AR) in a single environment to assist collaborations within multiple users. The concept combines advantages of both projector-based SAR for collaboration and HMD-based AR display in personalization to improve the efficiency of collaborative tasks. The presented concept is explored in a case study of industrial quality assurance scenario to show its effectiveness.

1 Introduction

Augmented Reality (AR) is a technology that integrates virtual objects into the real world [20]. It is the registration of projected computer-generated images over a user's view of the physical world. With this extra information presented to the user, the physical world can be enhanced or augmented beyond the user's normal experience. Additional information that is spatially located relative to the user can help to improve their understanding of the world in situ. AR interfaces enable people to interact with the real world in ways that are easily acceptable and understandable by users. For example, doctors can use the AR system to allow an intuitive real-time intraoperative orientation in image-guided interstitial brachytherapy [8], and to

J. Zhou (✉)
School of Computer and Information Science, University of South Australia, Australia
e-mail: Jianlong.Zhou@unisa.edu.au

L. Alem and W. Huang (eds.), *Recent Trends of Mobile Collaborative Augmented Reality Systems*, DOI 10.1007/978-1-4419-9845-3_13,
© Springer Science+Business Media, LLC 2011

Fig. 1 Example for a
screen-based video see-through
display. The locomotion of a
dinosaur is simulated over a
physical foot-print [3]

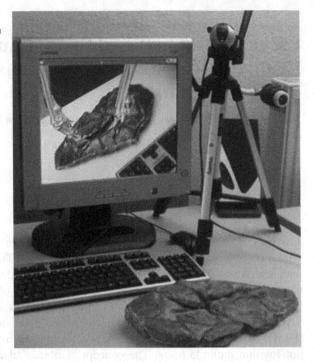

guide the liver thermal ablation in interventional radiology [14]. Doctors can also
use projector based AR for the intraoperative visualization of preoperatively defined
surgical planning data. The potential of AR for industrial processes is also increas-
ingly being investigated. However, the long tradition of AR systems has been based
on systems employing Head-Mounted-Displays (HMDs) that involve complex
tracking and complicated equipment worn by users. This can not meet industrial
requirements. Additional disadvantages such as limitations in Field-Of-View (FOV),
resolution, registration and bulkiness, make HMDs less attractive in industrial
applications [4].

Instead of body-attached displays, the emerging field of Spatial Augmented
Reality (SAR) detaches the technology from users and integrate it into the environment. SAR employs new display paradigms that exploit large spatially-aligned optical elements. Three different approaches exist in SAR, which mainly differ in the
way they augment the environment either using video see-through, optical see-
through or direct augmentation. Screen-based augmented reality makes use of
video-mixing (video see-through) and displays the merged images on a regular
screen (e.g. see Figure 1). Spatial optical see-through SAR generates images that
are aligned within the physical environment. Spatial optical combiners, such as planar or curved mirror beam splitters, transparent screens, or optical holograms are
essential components of such displays [4]. Projector-based SAR applies front-
projection to seamlessly project images directly on physical objects' surfaces,
instead of displaying them on an image plane (or surface) somewhere within the

Fig. 2 Projector-based augmentation of a large environment. Virtual model (upper left); Physical display environment (upper right); Augmented display (bottom) [10]

viewer's visual field [4]. Figure 2 is a projector-based augmentation of a large environment. Due to the decrease in cost and availability of projection technology, personal computers, and graphics hardware, SAR is now a viable option for use in an industrial setting. This chapter focuses on the projector-based SAR display and its applications in a collaborative environment.

This chapter firstly investigates SAR techniques and summarizes advantages and problems of SAR in presenting digital information to users. Specifically, the chapter outlines laser projector-based SAR in various applications, as we are investigating the use of laser projected SAR in the automotive industry. The chapter then presents a concept of portable collaborative SAR. The concept utilizes both projector-based SAR and HMD-based AR displays in a single environment to assist collaborations within multiple users. In this concept, a portable projector (e.g. laser projector) based SAR is used to project 3D digital information onto the physical object's surface. Meanwhile, each individual user utilizes an HMD-based AR to receive customized information about the position on the physical object's surface marked by the projector-based SAR. The concept combines advantages of both projector-based SAR and HMD-based AR display to improve the efficiency of collaborative tasks. The presented concept is used in a case study of industrial quality assurance scenario to show its effectiveness.

The chapter is organized as follows: Section 2 investigates advantages and problems of SAR, and presents typical applications of SAR in industries. Section 3 outlines laser projector-based SAR and shows that it partially solves problems of conventional projector-based SAR in presenting digital information to users. Section 4 presents a concept of portable collaborative SAR. This concept combines laser projector-based SAR and HMD-based AR display in a single environment to support collaborations within multiple users. Section 5 gives a case study which uses the presented concept in spot welding inspection in automobile industries. Finally, Section 6 summarizes and concludes the chapter.

2   Spatial Augmented Reality and Its Industrial Applications

SAR allows digital objects, images, and information to be added as real world artifacts by projecting onto surfaces in the environment with digital projectors. Bimber et al. [5] use conventional projectors that are calibrated in suitable locations to generate SAR scenes. They are able to show seemingly undistorted video and graphics on arbitrary surfaces in the environment, by means of pre-warping and color-adjusting the virtual data to counteract the reflection and perspective projection effects of the physical objects' surfaces. SAR benefits from the natural passive haptic affordances offered by physical objects [13]. This section outlines advantages and problems of SAR. Industrial applications of SAR systems are also discussed in this section.

2.1   Advantages of SAR

A key benefit of SAR is that the user is not required to wear a HMD and is therefore unencumbered by the technology. The user can physically touch the objects onto which virtual images are projected. In SAR, the FOV of the overall system is the natural FOV of the user, allowing him to use his peripheral vision. The range of SAR system's FOV can easily be extended by adding more projectors. Ultimately the FOV can emulate the full physical environment with a greater level of resolution with what is determined to be the correct number and position of projectors. Projector-based SAR allows possibly higher scalable resolution and bright images of virtual objects, text or fine details, than traditional HMD or handheld display solutions. Since virtual objects are typically rendered near their real-world locations, eye accommodation is easier [4, 18].

2.2   Problems of SAR

Like most of techniques, SAR also has some problems besides advantages in applications. The crucial problems with projector-based SAR are as follows [5, 18]:

- **Dependence on properties of display surfaces**. A light colored diffuse object with smooth geometry is ideal. Rendering vivid images on highly specular, low reflectance or dark surfaces, is practically impossible. The ambient lighting can also affect the contrast of images. This limits applications of SAR to controlled lighting environments with restrictions on the type of objects with which virtual objects will be registered.
- **Restrictions of the display area**. The display area is constrained to the size and shape of the physical objects' surfaces (for example, no graphics can be displayed beside the objects surfaces if no projection surface is present). Multi-projector configurations can only solve this problem if an appropriate display surface is present.
- **Shadow-casting**. Due to the utilisation of the front-projection, SAR has the problem of shadow-casting of the physical objects and of interacting users. This can be partially overcome by employing multiple projectors.
- **One active head-tracked user**. SAR also allows only one active head-tracked user at any instant in the environment because images are created in the physical environment rather than in the user's individual space. Time multiplexed shuttered glasses may be used to add more users that are active and head-tracked, but this requires the user to wear technology.
- **A single focal plane**. Conventional projectors only focus on a single focal plane located at a constant distance. It causes blur when projecting images onto non-planar surfaces. Multifocal projection technology [2] can solve this problem by employing multiple projectors.
- **Complexity of consistent geometric alignment and color calibration**. When the number of applied projectors increases, the complexity of consistent geometric alignment and color calibration is increased dramatically.

A major issue for SAR is the determination of suitable projection areas on the object itself. This limits the amount and complexity of information that can be presented [27]. Meanwhile, since the diffuse reflection is very small, only a minimal amount of light is reflected omni-directionally towards arbitrary viewer positions. Therefore, the projected and 3D aligned augmentations on the surface are not clear to viewers and have to be kept simple. So the main challenges include: How to project onto arbitrary surfaces; Where to mount the projectors; and How to provide adequate accuracy. Part of problems caused in conventional projector-based SAR can be finely solved by utilizing laser projectors as discussed in following sections.

2.3 SAR in Industrial Applications

AR technology was applied successfully in certain use cases in industries [19]. Several major application areas are identified: servicing and maintenance, design and development, production support, and training. Similarly, SAR systems have the potential to improve processes in a variety of application domains. For example, doctors could use the SAR to jointly visualize and discuss virtual information that is projected onto a patient or mannequin, while simultaneously visualizing remote collaborators whose imagery and voices are spatially integrated into the surrounding environment [5, 18].

In industries such as manufacturing, SAR could benefit a designer from the perceived ability to visually modify portions of a physically machined table-top model. The approach could also be used for product training or repair: one could set the product in the SAR environment and have the system render instructions directly on the product. Marner and Thomas [13] set up a SAR-based physical-virtual tool for industrial designers. The system simultaneously models both the physical and virtual worlds. SAR is then used to project visualizations onto the physical object, allowing the system to digitally replicate the designing process to produce a matching 3D virtual model. Olwal et al. [15] use SAR on industrial CNC-machines to provide operators with bright imagery and clear visibility of the tool and workpiece simultaneously. This helps to amplify the operator's understanding and simplify the machine's operation. Schwerdtfeger [24] uses HMD-based augmented reality to guide workers in a warehouse with pick information, which is named as pick-by-vision.

In the industry of automobiles, SAR can be used in quality assurance and maintenance as well as other applications. The quality assurance of spot welding is one of typical applications [26, 27] in automobile industries. In addition, SAR can also be used in job training in the automotive industry.

3   Laser Projector-Based SAR

Laser projector-based SAR has special properties compared with conventional projector-based SAR. This section shows related work on laser projectors and applications in SAR. Advantages of laser projectors are also discussed in this section.

3.1   Related Work

Wearable laser projectors have already been presented by Maeda et al. [12]. Kijima et. al [7] develop a hand-held laser projector to enhance the annotation of a barcode with additional information. Glossop and Wang [6] develop a laser projection AR system for computer-assisted surgery. The system uses rapidly scanned lasers to display information directly onto the patient. A well established industrial application of laser projector uses a table top system that mounts a laser projector to indicate directly on a circuit board where to place the next item [22, 27]. A laser can also be used in a remote collaboration AR system for annotation in the real workplace the user is focused on [9, 16].

There are other approaches toward using laser projectors in industrial applications. Zaeh and Vogl [28] use stationary laser projection to visualize tool trajectories and target coordinates in a robots environment by SAR technology. The system is

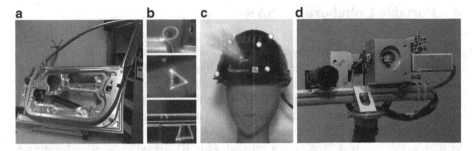

Fig. 3 (a) A white body of a car door, with about 50 welding points; (b) Projected images onto the white body; (c) First prototype of a laser projector: head-mounted, integrated into a helmet; (d) Second prototype: tripod-mounted, consisting of a wide angle camera for optical tracking, a galvanometer scanner and a 1 mW laser (from left to right) [27]

arranged and calibrated for a specific, static scene. The surfaces onto which information is projected may not be moved. MacIntyre and Wyvill [11] developed a laser projector that augments chickens in a processing line with automatically generated slaughter instructions. Schedwill and Scholles [23] develop a laser projection system for industrial uses, such as distance measurement and AR applications.

Schwerdtfeger et al. [25, 26, 27] set up an AR system that uses laser projectors. Figure 3 shows an example of this system. The system is used in the quality assurance of welding points. In this system, a hybrid information presentation approach is used: the laser projector is used to locate and display the position of welding points to be checked; an additional computer display is used to show complex what-to-do information to users. The system still requires users to read the computer display while focusing on the welding points, thus affects the work efficiency.

3.2　Advantages of Laser Projectors

The problems of the conventional projector-based SAR can be solved partially by the utilization of laser projectors. The laser projector-based SAR has following advantages:

- **Self calibrations**. The laser projector is self-calibrated and does not need additional calibrations compared with conventional projectors;
- **High bright laser beam**. The laser projector uses high bright laser beam and allows the user to perceive information from a large view and complex lighting conditions (even when viewed through an HMD);
- **Unlimited depth of focus**. The laser projector has "unlimited" depth of focus. The projected image on the object is in focus at any time.

Because of these advantages, information presentation systems using the laser projector based SAR are getting widely used in various applications, such as the aerospace industry.

4 Portable Collaborative SAR

In a natural face-to-face collaboration, people mainly use speech, gesture, gaze, and nonverbal cues to communicate. The surrounding physical world and objects also play an important role, particularly in a spatial collaboration tasks. Real objects support collaboration through their appearance, physical affordances, and ability to create reference frames for communication [1]. SAR technology promises to enhance such face-to-face communication. SAR interfaces blend the physical and virtual worlds, so real objects can interact with 3D digital content and improve users' shared understanding. Such interfaces naturally support face-to-face multi-user collaborations. In the collaborative SAR, co-located users experience a shared space that is filled with both real and virtual objects. Moreover, wearable computer based AR interface, such as HMD-based AR, in a collaboration makes the power of computer enhanced interaction and communication in the real world accessible any-time and everywhere [21]. HMD-based AR offers the flexibility for the personaliza-tion of information. HMD's allow for a customized AR visualization for each individual user in a collaboration.

This section presents a concept of portable collaborative SAR which utilizes both SAR and HMD-based AR to support collaborations within multiple users. In this concept, a portable projector (e.g. laser projector) based SAR is used to project global and positioning 3D digital information onto the physical object's surface. Meanwhile, each individual user utilizes an HMD-based AR to receive customized information about the position marked by the projector-based SAR on the physical object's surface. The portable laser projector and thus the portable collaborative SAR allow flexibility for locations of work places.

Figure 4 shows an illustration of this concept. The concept combines advantages of the projector-based SAR and the HMD-based AR at the same time: SAR projects digital marks onto the physical object's surface. These marks serve two purposes, 1) the provide a global physical world context for both user's to understand the prob-lem space and 2) a fiducial marker for the HMD-based AR visualizations. Various users who wear the HMD-based AR system can receive personalized information on the marked points.

As an example, Figure 5 shows that two operators view the same SAR projected digital mark (light blue disk), with personalised welding and inspection information displayed by the HMD screen for each individual operator. These light blue AR displayed graphical objects provide global overall information for both users. These can indicate graphically the position and type of weld. This acts a universal means of providing a "grounded" frame of reference for both users. Critical positions and information are physically common between the users, and therefore this removes any uncertainty due to tracking errors. The HMD AR displays are created using Tinmith Wearable Computer [17]. The operator who is an English speaker receives operation information in English, while the operator who is a Chinese speaker receives operation information in Chinese. In this figure, spot welding on a mechan-ical part are inspected. Each operator receives information on different aspects of

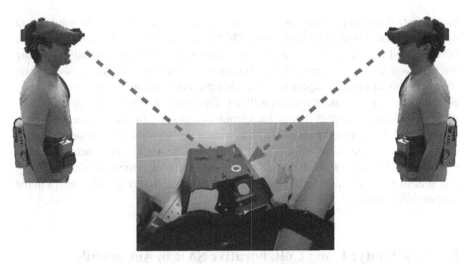

Fig. 4 Concept illustration of combining laser projector-based SAR and HMD-based AR for collaborations

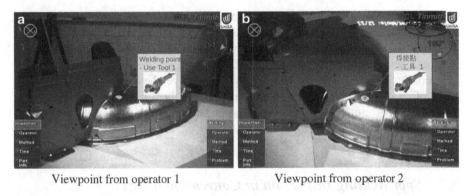

Viewpoint from operator 1   Viewpoint from operator 2

Fig. 5 Concept illustration of combining projector-based SAR and HMD-based AR visualisation for different operators

the welding spot. The light blue laser projected disk is treated as a tracking marker and global information. The marker-based AR is used to identify the marker as the reference point, and place the personalised instructions through the user's HMD according the reference point.

This concept has potential applications in various areas. First, it can be used in a collaborative training environment. The collaborative training environment allows trainees with different backgrounds (e.g. language) work together without additional helps. For example, a company has multiple factories in various countries. Operators doing the same task in different countires need to be trained. Users training in their own languages is preferred to improve the effectiveness of the training. Of course, operators can be classified into groups according to their languages. Multiple training sessions are then necessary for various groups based on languages. By using the collaborative training environment, one single training session is only used for all operators from different backgrounds. This is achieved by each trainee having a personalized interface while viewing the same target. The collaborative training environment not only improves effectiveness of the training session, but also allows trainees with different backgrounds to learn with each other in a single environment. Second, the concept can be used in military, diplomatic and other similar situations where participants have different backgrounds and need to collaborate with each other. Third, the concept benefits various industries in production support, designing and maintenance. Last but not least, similar to Chapter 11, the concept could be extended and used in remote collaborative tasks such as remote guiding, training and maintenance. Section 5 shows an example of using the presented concept in the automobile industry.

5   Case Study: Using Collaborative SAR in Automobile Industries

This section presents an application example that uses the presented concept in a training session of an industrial quality assurance scenario, where SAR is used to highlight spot welding to be inspected on an unpainted metal car part. The use of SAR can help operators to improve the efficiency of spot welding inspection in an automobile industry. The approach aims to remove the paper-based operation description sheet from operators' hands and relieve them from the heavy tiresome work, in order to improve the accurateness and efficiency of the inspection of spot welding.

5.1   Spot Welding Inspection in Conventional Ways

In the industry of automobiles, the quality of spot welding on car bodies needs to be inspected in regular intervals. For example, in an automobile company, a typical car has thousands of individual spot welds. In the process of making the vehicle, sub-assemblies are made and these assemblies have around 30–200 spot weldings. The spots have to be checked randomly from one to the next, even if the same type of part is checked — this has statistical reasons dealing with the occurrence of false negatives. Operators often do not check all spots on each body. They only check different certain number of spots on different bodies in a sequence. When all 200 spot welds are checked in a sequence, operators start a new spot sequence for checking. A variety of different methods are used to check spot welding: visual inspection, ultrasonic test, and destruction test.

The current procedure that operators use to check spot welding is as follows: the operator has a drawing of the testing body. The spots to be tested are marked in this drawing. First, the operator has to find the spot in the drawing. Then he has to find it on the body. After this, he has to choose the corresponding control method to finally perform the inspection. This manual inspection process has potential problems: the operator is easy to check wrong locations and wrong numbers of spot welding; it is also difficult for the operator to remember where to start and where to finish the checking on the checked body.

5.2 Using Laser Projector-Based SAR in Spot Welding Inspection

SAR benefits to spot welding inspection in the automobile industry. It facilitates presentation of projected digital AR information onto surfaces in structured work environments. Specifically, the portable laser projector-based SAR allows to project visual data onto arbitrary surfaces for the express purpose and providing just-in-time information to users in-situ within a physical work cell. It enables operators to do the spot welding inspection efficiently and effectively.

In this example, a laser projector mounted on a movable stand is employed to view and interact with digital information projected directly onto surfaces within a workspace. SAR provides guidance to operators of the next set of spot welding to inspect. The data items are projected onto the car body, providing instructions to operators. This removes the need to constantly refer to the instruction manual such as the operation description sheet, thus speeding up the operation and reducing errors.

The concept of portable collaborative SAR can be used in a training program of spot welding inspection. With the proposed technique, a typical training session can be conducted as follows: an instructor performs the spot welding inspection according to the digital marker pointed by laser projector-based SAR, and trainees wear HMDs to inspect operations. HMD-based AR allows trainees to access different information (e.g. welding method, welding time) of the same welding spot marked by the laser projector concurrently. For example, some trainees may want to know the welding method for the inspected spot and put this information on their HMD's screen, while others may want to display the inspection information in their own native languages. Because each trainee gets personalised information with HMD-based AR, each trainee can learn spot welding inspection from different aspects. This may improve the training efficiency for both the individual trainee and the overall training session.

There are benefits for providing in-situ data presentation for the spot welding inspection. First is the reduction in cognitive load of forcing people to remember specific tasks and the order they are required in. Second, the vehicles coming down the line are individually built (each car is different as they come down the line), and this requires unique information for each vehicle. Third, changes to the production information can be directly sent to the production line and displayed to the user. Last but not the least, it improves the inspection accuracy and efficiency greatly.

6 Conclusions

This chapter reviewed SAR techniques and summarized advantages and problems of SAR in presenting digital information to users. The chapter then presented a concept of portable collaborative SAR. The concept utilizes both projector-based SAR and HMD-based AR displays in a single environment to assist collaborations within multiple users. The concept combines advantages of both projector-based SAR for collaboration and HMD-based AR display in personalization to improve the efficiency of collaborative tasks. The presented concept was used in a case study of industrial quality assurance scenario to show its effectiveness.

Acknowledgements The authors would like to thank AutoCRC for the financial support in part and Thuong Hoang for setting up Tinmith Wearable Computer to produce Figure 5.

References

1. Billinghurst, M., Kato, H.: Collaborative augmented reality. Communications of the ACM **45**(7), 64–70 (2002)
2. Bimber, O., Emmerling, A.: Multifocal projection: A multiprojector technique for increasing focal depth. IEEE Transactions on Visualization and Computer Graphics **12**(4), 658–667 (2006)
3. Bimber, O., Gatesy, S.M., Witmer, L.M., Raskar, R., Encarnacão, L.M.: Merging fossil specimens with computer-generated information. IEEE Computer pp. 45–50 (2002)
4. Bimber, O., Raskar, R.: Modern approaches to augmented reality. In: SIGGRAPH'05: ACM SIGGRAPH 2005 Courses, p. 1. ACM (2005)
5. Bimber, O., Raskar, R.: Spatial Augmented Reality Merging Real and Virtual Worlds. A K Peters LTD (2005)
6. Glossop, N.D., Wang, Z.: Laser projection augmented reality system for computer-assisted surgery. International Congress Series **1256**, 65–71 (2003)
7. Kijima, R., Goto, T.: A light-weight annotation system using a miniature laser projector. In: Proceedings of IEEE conference on Virtual Reality 2006, pp. 45–46 (2006)
8. Krempien, R., Hoppe, H., Kahrs, L., Daeuber, S., Schorr, O., Eggers, G., Bischof, M., Munter, M.W., Debus, J., Harms, W.: Projector-based augmented reality for intuitive intraoperative guidance in image-guided 3d interstitial brachytherapy. International Journal of Radiation Oncology Biology Physics **70**(3), 944–952 (2008)
9. Kurata, T., Sakata, N., Kourogi, M., Kuzuoka, H., Billinghurst, M.: Remote collaboration using a shoulder-worn active camera/laser. In: Proceedings of IEEE International Symposium on Wearable Computers, pp. 62–69 (2004)
10. Low, K., Welch, G., Lastra, A., Fuchs, H.: Life-sized projector-based dioramas. In: Proceedings of Symposium on Virtual Reality Software and Technology (2001)
11. MacIntyre, B., Wyvill, C.: Augmented reality technology may bridge communication gap in poultry processing plants. http://gtresearchnews.gatech.edu/newsrelease/augmented.htm (2005)
12. Maeda, T., Ando, H.: Wearable scanning laser projector (wslp) for augmenting shared space. In: Proceedings of the 14th International Conference on Artificial Reality and Telexistence, pp. 277–282 (2004)
13. Marner, M.R., Thomas, B.H.: Augmented foam sculpting for capturing 3d models. In: Proceedings of IEEE Symposium on 3D User Interfaces (3DUI) 2010, pp. 63–70. Waltham, Massachusetts, USA (2010)

14. Nicolau, S., Pennec, X., Soler, L., Buy, X., Gangi, A., Ayache, N., Marescaux, J.: An augmented reality system for liver thermal ablation: Design and evaluation on clinical cases. Medical Image Analysis **13**(3), 494–506 (2009)

15. Olwal, A., Gustafsson, J., Lindfors, C.: Spatial augmented reality on industrial cnc-machines. In: Proceedings of SPIE 2008 Electronic Imaging, vol. 6804 (The Engineering Reality of Virtual Reality 2008). San Jose, CA, USA (2008)

16. Palmer, D., Adcock, M., Smith, J., Hutchins, M., Gunn, C., Stevenson, D., Taylor, K.: Annotating with light for remote guidance. In: Proceedings of the 19th Australasian conference on Computer-Human Interaction (OZCHI'07), pp. 103–110 (2007)

17. Piekarski, W., Thomas, B.H.: The tinmith system: Demonstrating new techniques for mobile augmented reality modeling. Journal of Research and Practice in Information Technology **2**(34), 82–96 (2002)

18. Raska, R., Welch, G., Fuchs, H.: Spatially augmented reality. In: Proceedings of IEEE and ACM IWAR'98 (1st International Workshop on Augmented Reality), pp. 11–20. San Francisco (1998)

19. Regenbrecht, H., Baratoff, G., Wilke, W.: Augmented reality projects in the automotive and aerospace industries. IEEE Computer Graphics and Applications **25**(6), 48–56 (2005)

20. Reiners, D., Stricker, D., Klinker, G., Mueller, S.: Augmented reality for construction tasks: Doorlock assembly. In: Proceedings of IEEE and ACM IWAR'98 (1st International Workshop on Augmented Reality), pp. 31–46. San Francisco (1998)

21. Reitmayr, G., Schmalstieg, D.: Mobile collaborative augmented reality. In: Proceedings of IEEE and ACM International Symposium on Augmented Reality, p. 114 (2001)

22. Royonic: http://www.royonic.com/en/ (2010)

23. Schedwill, I., Scholles, M.: Laser projection systems for industrial applications. http://www.ipms.fraunhofer.de/common/products/SAS/Systeme/laserprojmeasure-e.pdf (2008)

24. Schwerdtfeger, B.: Pick-by-vision: Bringing hmd-based augmented reality into the warehouse. Ph.D. thesis, Institut für Informatik der Technischen Universität München (2009)

25. Schwerdtfeger, B., Hofhauser, A., Klinker, G.: An augmented reality laser projector using marker-less tracking. In: Demonstration at 15th ACM Symposium on Virtual Reality Software and Technology (VRST'08) (2008)

26. Schwerdtfeger, B., Klinker, G.: Hybrid information presentation: Combining a portable augmented reality laser projector and a conventional computer display. In: Proceedings of 13th Eurographics Symposium on Virtual Environments, 10th Immersive Projection Technology Workshop(IPT-EGVE 2007) (2007)

27. Schwerdtfeger, B., Pustka, D., Hofhauser, A., Klinker, G.: Using laser projectors for augmented reality. In: Proceedings of the 2008 ACM symposium on Virtual reality software and technology (VRST'08), pp. 134–137 (2008)

28. Zaeh, M., Vogl, W.: Interactive laser-projection for programming industrial robots. In: Proceedings of the 5th IEEE and ACM International Symposium on Mixed and Augmented Reality (ISMAR'06), pp. 125–128 (2006)

Index

L. Alem and W. Huang (eds.), *Recent Trends of Mobile Collaborative
Augmented Reality Systems*, DOI 10.1007/978-1-4419-9845-3,
© Springer Science+Business Media, LLC 2011